Arrhythmia Induction in the EP Lab

Gabriel Cismaru
Editor

Arrhythmia Induction in the EP Lab

Springer

Editor
Gabriel Cismaru
Electrophysiology Laboratory, Rehabilitation Hospital
Iuliu Hatieganu University of Medicine and Pharmacy
5th Department of Internal Medicine, Cardiology-Rehabilitation
Cluj-Napoca
Romania

ISBN 978-3-319-92728-2 ISBN 978-3-319-92729-9 (eBook)
https://doi.org/10.1007/978-3-319-92729-9

Library of Congress Control Number: 2018968173

This Springer imprint is published by the registered company Springer Nature Switzerland AG
The registered company address is: Gewerbestrasse 11, 6330 Cham, Switzerland

Preface

The practice of electrophysiology requires a good understanding of intracardiac electrical signals, but beyond that it needs practical application of the theoretical principles. Induction of arrhythmia is the cornerstone of electrophysiology. Understanding the signals implies having the arrhythmia in front of the eyes, struggling to deduce the initiation, morphology, and timeline of the signals.

There have been remarkable advances in the mapping techniques, imaging, and catheters used for catheter ablation, but arrhythmia induction can be obtained by using the same old drugs described in this book.

How to Induce Arrhythmias was written to assist EP practitioners, fellows, and nurses in finding the doses and specific medications in the EP lab when the clinical arrhythmia cannot be induced by classical programed stimulation. In the first chapters, classical drugs are described: isoprenaline, orciprenaline, and atropine, as well as old drugs, used when the first two are unavailable—adrenaline, noradrenaline, dopamine, and dobutamine. Other various medications rarely used in the EP lab are also described because case reports showed their usefulness in clinical settings: aminophylline, adenosine, salbutamol, and ephedrine. Specific doses for children are shown in a chapter dedicated to pediatric electrophysiology. Two distinct chapters on energy drinks and caffeine were written because anecdotic cases reported sustained arrhythmias as side effects after their use.

The text is written in the same style for all chapters, in a simple way, following the same structure, with drawings, figures, and tables that assist easy finding of the information for a specific drug: all chapters start with an introduction, the chemical structure of the product, pharmacology, and mechanism of action; continue with doses, protocol of administration, side effects, and contraindications; and finish with the electrophysiological effects of the drug in animal and humans for different types of arrhythmias.

Electrophysiologists at different levels will find this book useful. Beginners will find out what medication can be used to facilitate arrhythmia induction, what is the sensibility and specificity of different drugs, how to administrate the medication, and what side effects to expect. EP practitioners will gain knowledge by reviewing the doses they might have long forgotten.

We would like to receive your critique and positive feedback on ResearchGate so that a new edition of *Arrhythmia Induction in the EP Lab* can improve to meet EP's needs.

Cluj-Napoca, Romania　　Gabriel Cismaru

Contents

About the Author

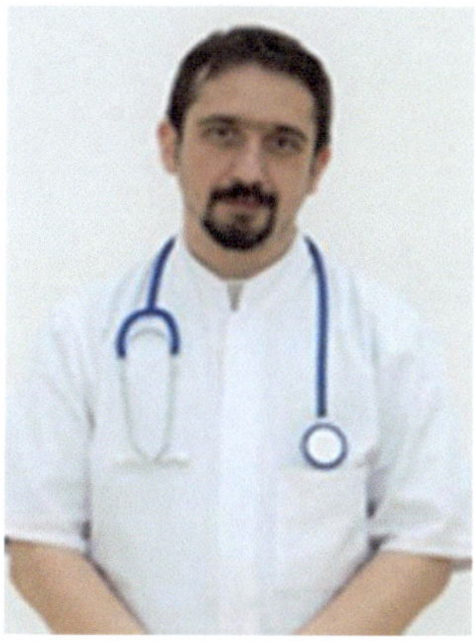

Gabriel Cismaru received his doctorate in medicine from the "Iuliu Hatieganu" University of Medicine and Pharmacy, Cluj-Napoca, in 2016, and earned his MD degree from the same university in 2005. After completing his residency in cardiology, Dr. Cismaru began his electrophysiology fellowship at the Institut Lorrain du Coeur et des Vaisseaux Louis Mathieu, Nancy, France. In 2011, he started to work at the Electrophysiology Laboratory of the Rehabilitation Hospital Cluj-Napoca. He has authored or coauthored peer-reviewed articles and book chapters in the field of electrophysiology and cardiac pacing.

The new advances in cardiac arrhythmia techniques increased the success rate of complete treatment for patients suffering from arrhythmias. However, the reproduction of the clinical arrhythmia might be difficult in hospital settings.

Arrhythmia Induction in the EP Lab is an extension of a PhD thesis in the field of cardiac electrophysiology. Various chapters describe methods to induce clinic arrhythmias, with a special chapter dedicated to arrhythmia induction in children. These topics are of great interest to electrophysiologists that deal with patients suffering from different types of arrhythmias.

Introduction: Why Do We Need Arrhythmia Induction?

1

Sorin Lazar

Abbreviations

AP	Accessory pathway
AT	Atrial tachycardia
AVNRT	Atrioventricular reentry (or reciprocating) tachycardia
AVRT	Atrioventricular reentrant (or reciprocating) tachycardia
EKG	Electrocardiogram
EP	Electrophysiology
ORT	Orthodromic reciprocating tachycardia
PVC	Premature ventricular contractions
Short RP tachycardia	Short R wave to P wave tachycardia
SVT	Supraventricular tachycardia
VT	Ventricular tachycardia

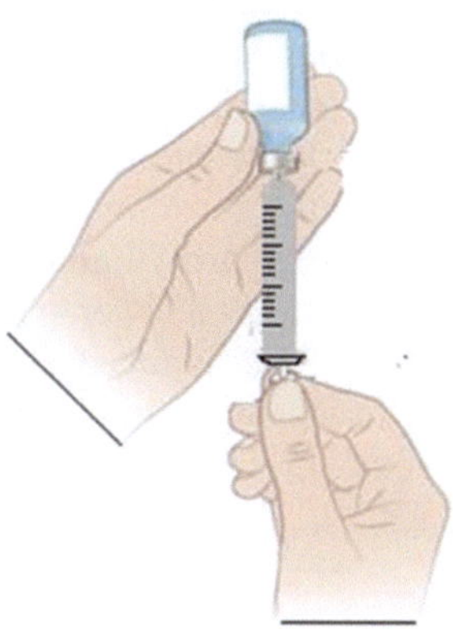

S. Lazar (✉)
Division of Cardiology, University of Illinois at Chicago, Chicago, IL, USA
e-mail: slazar1@uic.edu

G. Cismaru (ed.), *Arrhythmia Induction in the EP Lab*,
https://doi.org/10.1007/978-3-319-92729-9_1

1.1 Better Outcome of Catheter Ablation When Clinical Arrhythmia Is Inducible

One of the main challenges in evaluating a patient with palpitations is determining the type of arrhythmia that is responsible for the patient's symptoms. Many of the available outpatient monitors have limitations in the number of channels they can record or the amount of data they can store for analysis. In many instances, the initiation and termination of the arrhythmia are not captured on these monitors, and essential information for a correct diagnosis is not available. Ideally, the arrhythmia should be recorded on a 12-lead electrocardiogram for a correct diagnosis, but most of the time, this is possible only when the arrhythmia is sustained long enough to be still present when the patient arrives at the hospital. In the best case scenario, the arrhythmia is recorded on a heart monitor or 12-lead EKG, and then a pre-ablation strategy can be developed. Depending on the protocol used to induce arrhythmia, the clinical arrhythmia might be inducible along with other arrhythmias which might not be clinical. Knowing the type of arrhythmia the patient has as outpatient helps guide the ablation of the inducible arrhythmia that has similar characteristics, rather than map and ablate all inducible arrhythmias in the EP lab.

One example of this is a patient with palpitations and WPW at baseline, which suggest possible AVRT, but during the EP study, either the accessory pathway (AP) has an ERP that would make AVRT unlikely, with the clinical arrhythmia being AT or AVNRT and AP being a bystander [1].

Another example is PVCs with morphology suggesting outflow tract origin, but during the EP study with isoproterenol infusion, multiple PVC morphologies originating from other areas of the heart are induced. The target of the ablation is usually the morphology suggested by the outpatient recording.

For arrhythmias identified by event recorders only, where many times we don't have the initiation or termination of the tachycardia, the cycle length of the inducible tachycardia should be close to the outpatient one, although the specificity of this finding might be low due to different physiologic states between outpatient and during the electrophysiological study, where the patient might be sedated and undergoing drug infusions [2–4]. In the situations when we don't have a recorded clinical arrhythmia, then inducing an arrhythmia that reproduces the patient symptoms might be useful, although it has a lower specificity. Most of the time, in this situation, ablation of the arrhythmia induced during the EP study might be useful [5]. Special mentioning needs to be made of the patients who have an implantable device, where intracardiac electrograms are very useful in determining the mechanism of arrhythmia.

1.2 Catheter Ablation of Accessory Pathways During ORT and AVNRT

There are different strategies to map and ablate an AP [6]. If there is evidence that the AP is involved in the tachycardia, then ablation in sinus rhythm can be performed either by antegrade mapping or retrograde mapping during ventricular pacing. The end goal of the ablation is elimination of preexcitation. Alternatively, AP mapping can be performed during sustained SVT. If the arrhythmia is proven to be AVRT (atrioventricular reentrant tachycardia) and if the mechanism of the arrhythmia is ORT (orthodromic reentrant tachycardia), then the earliest retrograde atrial activation during sustained SVT is targeted. If the AVRT is antidromic, the earliest ventricular activation is mapped to determine the ventricular insertion of the AP. Ablation of the AP in sinus rhythm without being able to induce SVT carries the risk that the AP might be a bystander, and the only way to prove or disprove this is during sustained tachycardia. For this reason, it is mandatory to induce the arrhythmia for a correct diagnosis, especially if there is no documentation of the clinical arrhythmia on 12-lead EKG or cardiac monitor. Every effort should be made to induce arrhythmia to prove that AP is involved in the arrhythmia mechanism.

If the clinical suspicion of AVNRT (atrioventricular node reentrant tachycardia) is high despite non-inducibility of arrhythmia, slow pathway modification could be performed, but the outcomes are less favorable than when SVT is inducible in the EP lab. In the absence of inducible arrhythmia, some operators might hesitate to perform empirical ablation of the slow pathway, as there is no clear procedural end point and due to the risk of AV block [7, 8]. For patients with inducible AVNRT, the overall procedural success approaches 96% both for cryotherapy and for radiofrequency catheter ablation [9]. On the contrary, Shurrab et al. [10] showed that the success rate for patients with non-inducible AVNRT was 83.7% at 17 months, and this might be due to inability to determine if the mechanism of the clinical short RP tachycardia, suggesting AVNRT, is not actually AT, with an AV delay close to the tachycardia cycle length. Therefore every attempt should be made to induce AVNRT in the EP lab using isoprenaline, adrenaline, atropine, adenosine, etc. to create a perfect balance between the slow and fast pathway conductions to facilitate arrhythmia induction to be able to differentiate between the two arrhythmias.

1.3 Activation Mapping in Patients with Premature Ventricular Contractions

The success of a PVC ablation is dependent on the frequency of the PVC during the procedure. There are different ways to correctly identify the source of the PVC. One of the most commonly used method in the EP laboratory is activation mapping, when using a three-dimensional mapping system, the site of earliest activation during PVC is mapped and targeted. Unfortunately, many of the PVCs are adrenergic dependent, and with minimal sedation required for the procedure, the frequency of PVCs decreases to the point where activation mapping is almost impossible. In this situation, a few monomorphic PVCs are enough to create a template, and then pacemaps are correlated with the template to determine the origin of the PVC. The drawback of this method is that similar pacemaps are correlating reasonably well over wide areas of the ventricle, making it a less preferred method of mapping. For a PVC ablation, inducibility and activation mapping (either spontaneous or during the drug infusion) are essential for a good long-term outcome [11, 12].

1.4 Activation Mapping in Patients with Ventricular Tachycardia

Similar with PVC mapping, VT mapping can be performed using pacemaps, with similar drawback. This method is preferred only when VT is hemodynamically unstable. Stevenson et al. published the algorithm of determining the critical isthmus regions in hemodynamically stable ischemic VT during sustained arrhythmia [13]. During reentrant VT, using entrainment mapping, the authors showed that the VT exit sites can be localized precisely, and RF application in the area successfully terminates the VT. For a successful ablation of a focal VT, inducibility of the arrhythmia is essential for a good long-term outcome. Most of the time, we have to use multiple-drug infusions, in escalating doses, to induce clinical arrhythmia.

1.5 Reproducibility of Symptoms During Electrophysiological Testing

When the outpatient arrhythmia documentation is not available, the only way to determine if the arrhythmia we induced in the EP lab is the clinical one is to determine if the symptoms the patient is experiencing during induced arrhythmia are correlating with the patient's symptoms as outpatient. This method though has a very low sensitivity and specificity.

References

1. Smith WM, Broughton A, Reiter MJ, Benson DW Jr, Grant AO, Gallagher JJ. Bystander accessory pathway during AV node re-entrant tachycardia. Pacing Clin Electrophysiol. 1983;6(3 Pt 1):537–47.
2. Niksch A, Liberman L, Clapcich A, Schwartzenberger JC, Silver ES, Pass RH. Effects of remifentanil anesthesia on cardiac electrophysiologic properties in children undergoing catheter ablation of supraventricular tachycardia. Pediatr Cardiol. 2010;31(7):1079–82.
3. Pires LA, Huang SK, Wagshal AG, Kulkami RS. Electrophysiological effects of propofol on the normal cardiac conduction system. Cardiology. 1996;87(4):319–24.
4. Hermann R, Vetterman J. Change of ectopic supraventricular tachycardia to sinus rhythm during administration of propofol. Anesth Analg. 1992;75(6):1030–2.
5. Tsiachris D, Koutagar J, Gatzoulis KA, Arsenos P, Rigatou A, Dilaveris P, et al. Diagnosis and management of phantom tachycardias based on an electrophysiologically guided approach. Hell J Cardiol. 2016;57(5):340–4.
6. Jackman WM, Wang XZ, Friday KJ, Roman CA, Moulton KP, Beckman KJ, et al. Catheter ablation of accessory atrio-ventricular pathways (Wolff-Parkinson-White) by radiofrequency current. NEJM. 1991;324(23):1605–11.
7. Hanninen M, Yeung-Lai-Wah N, Massel D, Gula LJ, Skanes AC, Yee R, et al. Cryoablation versus RF ablation for AVNRT: a meta-analysis and systematic review. J Cardiovasc Electrophysiol. 2013;24(12):1354–60.
8. Liao JN, Hu YF, Wu TJ, Fong AN, Lin WS, Lin WJ, et al. Permanent pacemaker implantation for late atrioventricular block in patients receiving catheter ablation for atrioventricular nodal reentrant tachycardia. Am J Cardiol. 2013;111(4):569–73.

9. Schwagten B, Knops P, Janse P, Kimam J, Van Belle Y, Szill-Torok T, et al. Long-term follow-up after catheter ablation for atrioventricular nodal reentrant tachycardia: a comparison of cryothermal and radiofrequency energy in a large series of patients. J Interv Card Electrophysiol. 2011;30(1):55–61.
10. Shurrab M, Szill-Torok T, Akca F, Tlong I, Kagal D, Newman D, et al. Empiric slow pathway ablation in non-inducible supraventricular tachycardia. Int J Cardiol. 2015;179:417–20.
11. Azegami K, Wilber DJ, Arruda M, Lin AC, Denman RA. Spatial resolution of pacemapping and activation mapping in patients with idiopathic right ventricular outflow tachycardia. J Cardiovasc Electrophysiol. 2005;16(8):823–9.
12. Dixit S, Gerstenfeld EP, Callans DJ, Marchlinski FE. Electrocardiographic patterns of superior right ventricular outflow tract tachycardias: distinguishing septal and free-wall sites of origin. J Cardiovasc Electrophysiol. 2003;14(1):1–7.
13. Stevenson WG, Friedman PL, Sager PT, Saxon LA, Kokovic D, Harada T, et al. Exploring postinfarction reentrant ventricular tachycardia with entrainment mapping. JACC. 1997;29(6):1180–9.

How to Induce Arrhythmias by Atrial and Ventricular Programmed Stimulation?

2

Celestino Sardu, Valerio Giordano, Antongiulio Donatiello, Raffaele Marfella, Giuseppe Paolisso, Maria Rosaria Rizzo, and Michelangela Barbieri

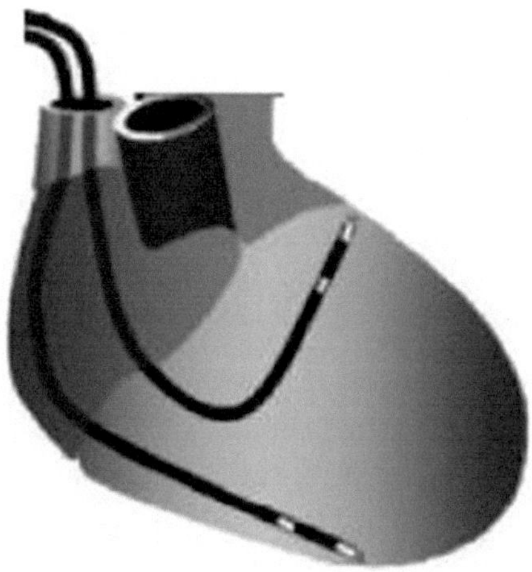

C. Sardu (✉)
Department of Medical, Surgical, Neurological, Metabolic and Aging Sciences, Universita degli Studi della Campania Luigi Vanviteli, Caserta, Italy
e-mail: celestino.sardu@unicampania.it

V. Giordano
Department of Arrhythmias and Cardiovascular Diseases, Clinic Center "Nostra Signora di Lourdes" Hospital, Naples, Italy

A. Donatiello
FTE AF Division, Abbott, Milan, Italy
e-mail: adonatiello@sjm.com

R. Marfella · G. Paolisso · M. R. Rizzo · M. Barbieri
Department of Medical, Surgical, Neurological, Metabolic and Aging Sciences, University of Campania "Luigi Vanvitelli", Naples, Italy
e-mail: raffaele.marfella@unicampania.it; giuseppe.paolisso@unicampania.it; mariarosaria.rizzo@unicampania.it; michelangela.barbieri@unicampania.it

G. Cismaru (ed.), *Arrhythmia Induction in the EP Lab*,
https://doi.org/10.1007/978-3-319-92729-9_2

2.1 Introduction

In recent decades, invasive electrophysiological study (ES) has become an important instrument to evaluate patients with conduction disturbance and cardiac arrhythmias [1]. Using catheters placed in heart chambers via central vein and/or central arterial access, ES may evaluate sinus node function, atrioventricular conduction, and tachyarrhythmias. Cardiac arrhythmias may not always be present in the baseline condition. Therefore, it is necessary to induce these arrhythmias by programmed pacing protocols [1]. Commonly arrhythmias are seen as chaotic alterations of the normal heart conduction and of the normal heart rhythm, and then they are defined as cardiac rhythm disorders [2]. Cardiac arrhythmias present with a common phenotype, characterized by irregularity of the cardiac rhythm and related clinical symptoms [2]. In this setting, ES may be performed by using different diagnostic and pacing catheters and pacing protocols. Programmed pacing protocols involve incremental pacing, coupled with the introduction of single or multiple premature stimuli during one or more drive cycles [3]. The pacing protocols are performed with a current output of twice the diastolic threshold or more, and at one or more sites [3]. Therefore, a great discrepancy may exist between arrhythmia induction techniques in different laboratories. However, in the majority of cases, arrhythmias are due to specific arrhythmic electrical, anatomical, and/or electroanatomical circuits [2]. These circuits respond to specific conduction properties of the systolic and diastolic electrical phases, which are reproducible and evocable by external triggers and by specific pacing techniques [2]. Moreover, in the light of these observations, we have to stress the concept that induction and stimulation programs have to be selected and then paced to test the arrhythmic circuits for refractoriness and to trigger the conduction properties of the arrhythmic pathways. Therefore, we have to make arrhythmia induction protocols more uniform and as standardized as possible to avoid all possible bias. Indeed, how to induce arrhythmias by atrial and ventricular stimulation remains a relevant question that needs a specific and unique response. To respond to this question, we would like to introduce the concept that a pacing protocol to induce cardiac arrhythmias may be standard and programmed [2]. As first, by programmed pacing, physicians may study the properties of the cardiac conduction system. This may be secondarily achieved by introduction of early stimuli to determine the conduction response [2], as a specific arrhythmia induction protocol. As discussed earlier, the type of induction and the chosen programmed stimulation protocol may be selected with regard to the type of arrhythmia the patient is suspected to have. In fact, re-entry tachycardias may usually be triggered using extrastimuli to stimulate the conduction pathways in slow conduction and fast conduction ways [2]. Differently, automatic tachycardias not due to re-entry mechanisms may be more easily induced by burst pacing [2]. In this chapter we would like to introduce pacing protocols to induce cardiac arrhythmias. Apart from the similarity of the diagnostic and pacing catheters, and in the setting of programmed stimulation protocols, the differences in the paced heart chambers and in the induced tachycardias teach us to separate the discussion of atrial stimulation from ventricular stimulation protocols. Therefore, we schematically divide the arrhythmia induction protocols into two separate chapter sections, discussing atrial stimulation protocols and ventricular stimulation protocols.

2.2 Atrial Stimulation

To perform atrial stimulation, the authors place diagnostic quadripolar and/or decapolar catheters in the atrial chambers. These catheters are introduced by central vein access, to map and to pace the right atrium appendage, the coronary sinus, and along the tricuspid valve annulus as indicated by a radioscopic biplane view of the heart chambers. Sometimes, pacing maneuvers may be performed by direct access to the left atrium, at the authors' discretion. To induce atrial arrhythmias the authors perform programmed pacing protocols divided into coupled pacing protocols and a burst pacing protocol [2]. In the case of a programmed coupled pacing protocol, the authors set a standard pacing protocol choosing a drive of 8 beats as an S1 interval of 600, 500, and then 400 ms, coupled with a first extrastimulus, called S2, that is conventionally at least 60 ms higher than the documented Wenckebach interval [2] (Fig. 2.1). Normally, the authors start with an S1 of 600 ms, then decreasing to 500 and 400 ms. The choice of the S1 cycle length may also depend on the patient's heart rate. Therefore, at the authors' discretion the coupled S2 interval is decreased by 10–20 ms for each new pacing train, in a manual and/or automatic way [2]. During each pacing train it is relevant to observe a resting period of 4 s, and to register and to note every arrhythmic event that occurs. In the case of atrial stimulation of re-entrant arrhythmias we may assist with an increase in the supra-Hisian (AH) interval by at least 50 ms from one train to the next, and this is called an "AH jump" [4]. This phenomenon is due to the pacing of the slow pathway of a re-entrant

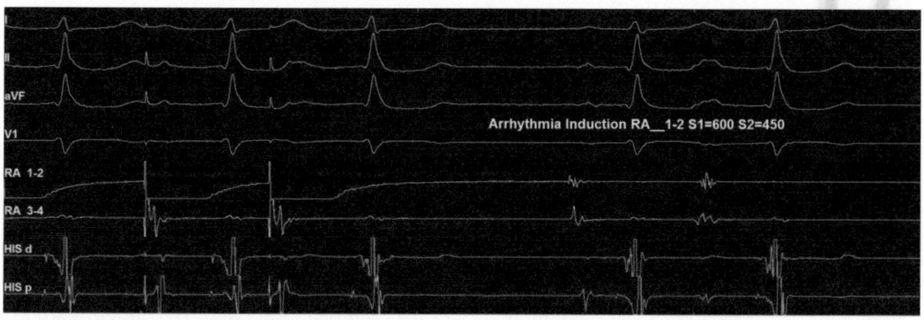

Fig. 2.1 Representation of programmed coupled atrial pacing for atrial arrhythmia induction. The drive of 8 beats is indicated as S1, at a cycle length of 600 ms. The coupled extrastimulus (S2) is at an interval of 450 ms. The pulse amplitude is 10 mA; the duration is 2 ms. The *upper part of the figure* shows the DI, DII, aVF, and V1 surface ECG derivations. The *yellow color* denotes the right atrium (RA: 1–2 is distal, 3–4 is proximal); the *green color* denotes the His bundle (HIS: d is distal, p is proximal). In this case the authors prefer to use quadripolar diagnostic catheters. (Image created by C. Sardu, WorkMate Claris polygraph, Abbot)

arrhythmic circuit, during the pass on a slow conduction pathway, indicating dual AV node physiology [4]. During programmed and coupled atrial pacing, we may reach a stimulation interval where the atrial pacing is not conducted through the atrioventricular node to the ventricles. This stimulation interval is called the atrioventricular node effective refractory period (AVNERP) [2] (Fig. 2.2). To induce atrial arrhythmias during programmed pacing, we start to shorten the S2 interval until the pacing signal no longer causes the atrium to contract, reaching the atrial effective refractory period (AERP) [2]. Therefore, reaching the refractory S2 interval, we increase the S2 interval to at least 20 ms above the AVNERP, and we repeat this coupled programmed stimulation introducing the secondary (S3) and then the third (S4) coupled extrastimulus as discussed earlier for S2 [2] and/or the AVNERP (in the case of re-entrant atrial arrhythmia induction). Reaching triple refractoriness of the coupled extrastimuli (S2–S3–S4), we stop the S1 programmed stimulation, switching the S1 interval from 600 to 500 and then 400 ms [2]. Moreover, we repeat the same induction protocol until S4 refractoriness occurs and/or in the case of tachycardia induction [2] (Fig. 2.3). Completing all this programmed pacing protocol from 600 to 400 ms in the S1 interval, and until refractoriness of the triple coupled extrastimulus (S2–S3–S4) occurs, we may start burst pacing to achieve a clinical arrhythmia and especially in the case of clinical atrial tachycardia induction [5] (Fig. 2.4). This second type of programmed pacing modality is different from coupled pacing for atrial arrhythmias due to re-entry circuits, and is related to continuous atrial pacing from 300 ms and down by 10–20 ms until 200 ms [5]. In the case of atrial burst pacing there are a few rules to follow. First, the authors consider

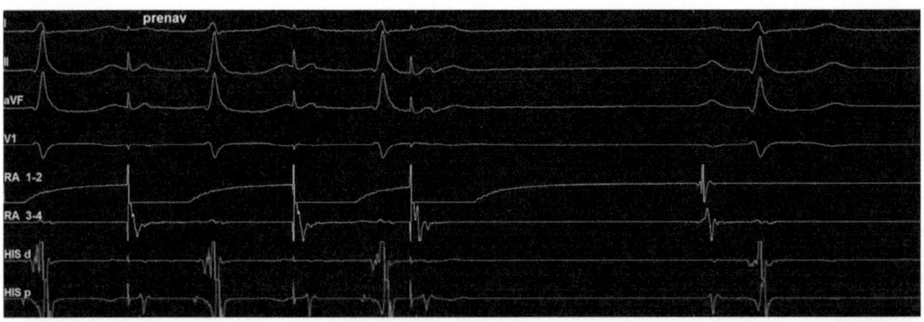

Fig. 2.2 Representation of programmed coupled atrial pacing to evaluate the atrioventricular node effective refractory period (AVNERP). The drive of 8 beats is indicated as S1, at a cycle length of 600 ms. The coupled extrastimulus (S2) is at an interval of 400 ms. The pulse amplitude is 10 mA; the duration is 2 ms. As can be seen, we may reach the stimulation interval where the pacing interval does not conduct through the AV node to the ventricles, then called the AVNERP. The *upper part of the figure* shows the DI, DII, aVF, and V1 surface ECG derivations. The *yellow color* denotes the right atrium (RA: 1–2 is distal, 3–4 is proximal); the *green color* denotes the His bundle (HIS: d is distal, p is proximal). In this case the authors prefer to use quadripolar diagnostic catheters. (Image created by C. Sardu, WorkMate Claris polygraph, Abbot)

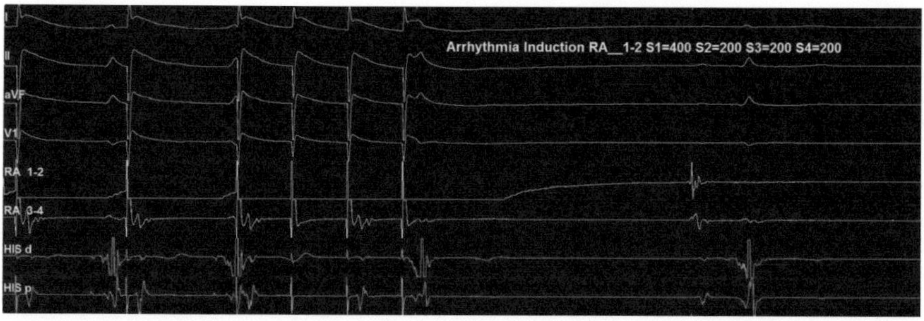

Fig. 2.3 Representation of programmed coupled atrial pacing for atrial arrhythmia induction. The drive of 8 beats is indicated as S1, at a cycle length of 400 ms. The coupled extrastimuli (S2, S3, and S4) are at an interval of 200 ms. The *upper part of the figure* shows the DI, DII, aVF, and V1 surface ECG derivations. The *yellow color* denotes the right atrium (RA: 1–2 is distal, 3–4 is proximal); the *green color* denotes the His bundle (HIS: d is distal, p is proximal). In this case the authors prefer to use quadripolar diagnostic catheters. (Image created by C. Sardu, WorkMate Claris polygraph, Abbot)

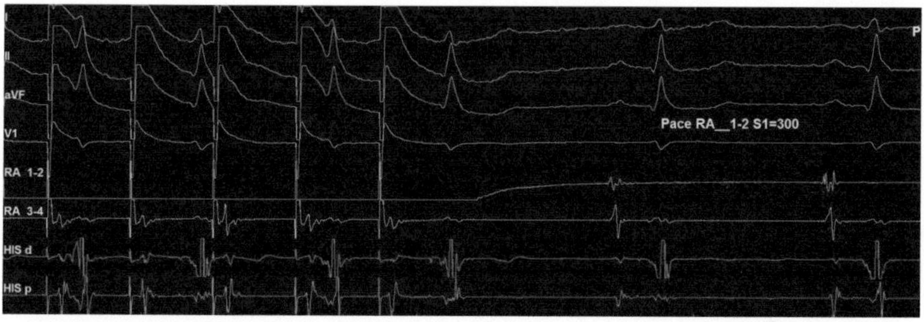

Fig. 2.4 Representation of programmed burst atrial pacing for atrial arrhythmia induction. The drive of beats is indicated as S1, at a cycle length of 300 ms. The *upper part of the figure* shows the DI, DII, aVF, and V1 surface ECG derivations. The pulse amplitude is 10 mA; the duration is 2 ms. The *yellow color* denotes the right atrium (RA: 1–2 is distal, 3–4 is proximal); the *green color* denotes the His bundle (HIS: d is distal, p is proximal). In this case the authors prefer to use quadripolar diagnostic catheters. (Image created by C. Sardu, WorkMate Claris polygraph, Abbot)

atrial burst pacing below 200 ms to be contraindicated [5]. Second, it seems intuitive and is commonly agreed to turn off the stimulator immediately in the case of atrial arrhythmia induction [6] (Fig. 2.5). During these pacing protocols we have to choose a standard value for the pacing pulse amplitude in milliamperes, and a duration in milliseconds. Normally these values have to be calculated and then chosen as the lowest values for atrial local capture [2]. All these pacing protocols are started in baseline conditions and may be repeated without coexisting contraindications during infusion of drugs interfering with vagal and sympathetic tone, and during patient maneuvers such as a hand grip [2].

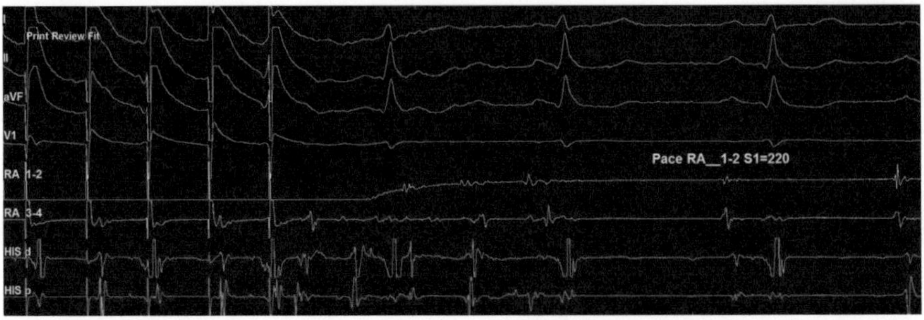

Fig. 2.5 Representation of programmed burst atrial pacing for atrial arrhythmia induction. As indicated in the text, the authors perform gradually increasing pacing from 300 to 200 ms, and/or until induction of an atrial arrhythmia. In this case the drive of beats is indicated as S1, and the authors (as discussed in the text) stop the pacing at a cycle length of 220 ms because nonsustained atrial tachycardia is induced. The *upper part of the figure* shows the DI, DII, aVF, and V1 surface ECG derivations. The pulse amplitude is 10 mA; the duration is 2 ms. The *yellow color* denotes the right atrium (RA: 1–2 is distal, 3–4 is proximal); the *green color* denotes the His bundle (HIS: d is distal, p is proximal). In this case the authors prefer to use quadripolar diagnostic catheters. (Image created by C. Sardu, WorkMate Claris polygraph, Abbot)

2.3 Ventricular Stimulation

To perform ventricular stimulation the authors set a programmed pacing protocol with a catheter placed in the ventricular chambers. The choice of the catheter and the catheter position may differ between authors. Commonly, the authors prefer to perform quadripolar mapping and place a pacing catheter in at least two different right ventricular positions, such as the right ventricular apex and right ventricular outflow tract, as indicated by biplane fluoroscopic imaging of the heart chambers. Uncommonly, ventricular stimulation may be performed by left ventricular access. During ventricular pacing we study retrograde conduction of the atrioventricular node, and/or we may stimulate a concealed accessory pathway, and this may be the first relevant observation during an ES [2]. In fact, ventricular pacing may be used first to study retrograde atrioventricular node conduction, and secondarily to induce suspected arrhythmias [2]. Therefore, during ventricular pacing, and in the case of a suspected supraventricular arrhythmia, it is important to map the left atrium and left ventricle conduction, introducing a secondary diagnostic catheter into the coronary sinus (a few authors also use a third catheter placed in the right atrial appendage) to observe the retrograde atrioventricular node activation [2] (Fig. 2.6). In fact, during ventricular pacing we may observe concentric and decremental retrograde atrioventricular node conduction in the case of retrograde normal atrioventricular node activation [2]. Sometimes, during S1 ventricular pacing we may observe retrograde atrioventricular dissociation, which confirms retrograde normal atrioventricular node activation [2]. On the other hand, we may not see decremental retrograde atrioventricular node conduction, which may also be eccentric (not retroactivated as proximal to distal by local coronary sinus electrogram analysis), by concealed retrograde atrioventricular accessory pathway pacing [2] (Fig. 2.7). Therefore, during

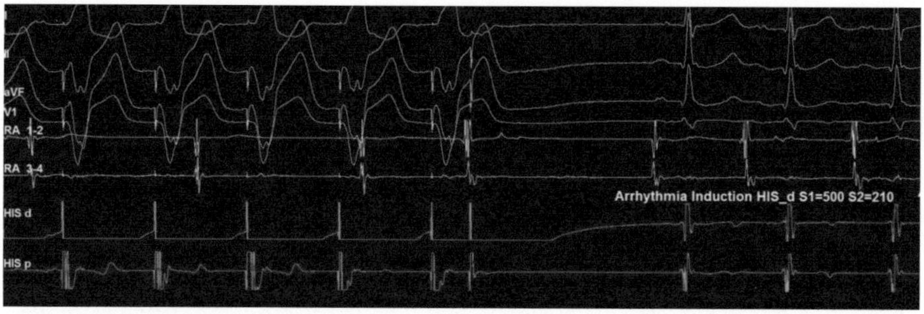

Fig. 2.6 Representation of programmed and coupled ventricular pacing to study retrograde atrioventricular node conduction. As indicated in the text, the authors perform 8-beat drive pacing as S1 (in this case 500 ms) coupled with gradually increasing pacing as S2 (in this case S2 is 210 ms). It can be seen that S2 is not conducted to the right ventricle, reaching the effective refractory period of the right ventricle. More than this, retrograde normal nodal conduction can be seen as atrioventricular retrograde dissociation during the drive pacing. The *upper part of the figure* shows the DI, DII, aVF, and V1 surface ECG derivations. The pulse amplitude is 10 mA; the duration is 2 ms. The *yellow color* denotes the right atrium (RA: 1–2 is distal, 3–4 is proximal); the *green color* denotes the His bundle (HIS: d is distal, p is proximal), which in this image is placed in the right ventricular apex. In this case the authors prefer to use quadripolar diagnostic catheters. (Image created by C. Sardu, WorkMate Claris polygraph, Abbot)

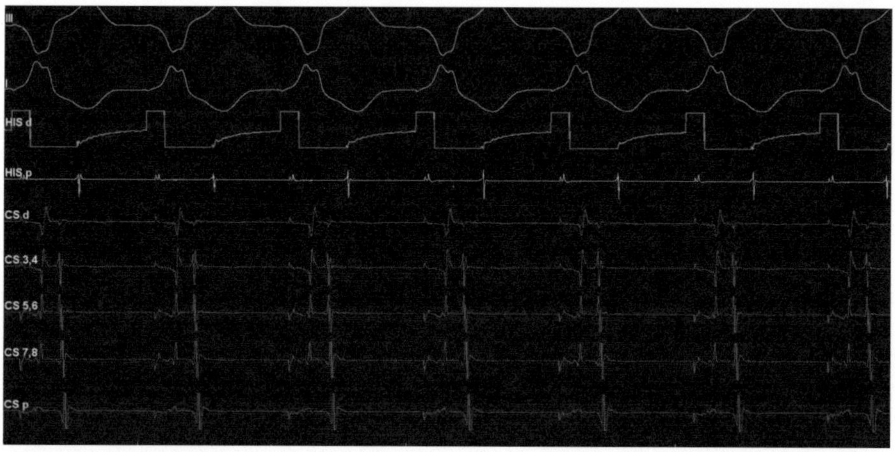

Fig. 2.7 Representation of programmed and coupled ventricular pacing to study retrograde atrioventricular node conduction. As indicated in the text, the authors perform 8-beat drive pacing as S1 (in this case 600 ms) coupled with gradually increasing pacing until retrograde atrioventricular refractoriness occurs, as S2 (in this case S2 is 350 ms). The figure is stopped at the S1 drive because during S1 pacing, retrograde eccentric nodal conduction occurs via a concealed retrograde bypass tract. The *upper part of the figure* shows the DI, DII, aVF, and V1 surface ECG derivations. The pulse amplitude is 10 mA; the duration is 2 ms. The *yellow color* denotes the His bundle (HIS: 1–2 is distal, 3–4 is proximal) placed in the right ventricular apex; the *red color* denotes the coronary sinus (CS: d is distal, p is proximal), in this case placed in the CS ostium. In this case the authors prefer to use a quadripolar diagnostic catheter for HIS and a decapolar catheter for CS (discussed in the text). (Image created by C. Sardu, WorkMate Claris polygraph, Abbot)

ventricular pacing, we may perform a pacing protocol to study retrograde atrioventricular node activation, and ventricular pacing to induce supraventricular and/or ventricular arrhythmias [2]. As mentioned earlier, we may perform induction pacing for ventricular arrhythmias by a programmed pacing protocol similar to that used for atrial arrhythmia induction. Moreover, we may perform coupled stimulation with eight pacing trains in the drive as S1 with progressive extrastimuli until the refractory periods are found, and/or burst pacing [7] (Figs. 2.8, 2.9, 2.10, and 2.11). During coupled pacing, one may reach the retrograde atrioventricular node effective refractory period (RAVNERP) as an interval in which ventricular pacing is not followed by a retroconducted atrial signal, and then the ventricle effective refractory period (VERP) as an interval in which the ventricular paced extrastimulus is not followed by a local ventricular electrogram [7]. The principal scope of ventricular stimulation is to induce ventricular arrhythmias by programmed pacing protocols. These pacing protocols may be coupled pacing protocols and/or a burst pacing protocol, and the choice of the correct pacing protocol to perform may represent the most relevant question in the setting of life-threatening ventricular arrhythmia induction [8]. As part of this, sometimes ventricular pacing may induce several types of supraventricular tachycardia such as atrioventricular node re-entry, atrioventricular re-entry, and even atrial fibrillation. Generally, to induce ventricular tachycardias the authors perform ventricular pacing by programmed coupled ventricular pacing, by a drive of 8 beats at an S1 interval of 600, 500, and then 400 ms (sometimes the authors prefer only two different drives), decreasing the extrastimuli by 10–20 ms for each new pacing train, in a manual and/or automatic way [7]. The extrastimuli are coupled with the pacing interval until ventricular conduction

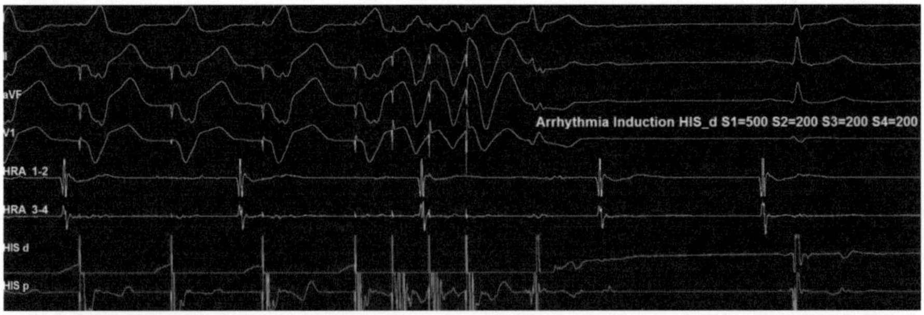

Fig. 2.8 Representation of programmed and coupled ventricular pacing to induce ventricular arrhythmias. As indicated in the text, the authors perform 8-beat drive pacing as S1 (in this case 500 ms) coupled with gradually increasing pacing by triple coupled beats as S2, S3, and S4 until refractoriness occurs (in this case S2, S3, and S4 are 200 ms). The *upper part of the figure* shows the DI, DII, aVF, and V1 surface ECG derivations. The pulse amplitude is 5 mA; the duration is 2 ms (the refractory ventricular conduction is at 2 mA with a duration of 2 ms). The *yellow color* denotes the right atrium (RA: 1–2 is distal, 3–4 is proximal); the *green color* denotes the His bundle (HIS: d is distal, p is proximal), which in this image is in the right ventricular apex. In this case the authors prefer to use quadripolar diagnostic catheters. (Image created by C. Sardu, WorkMate Claris polygraph, Abbot)

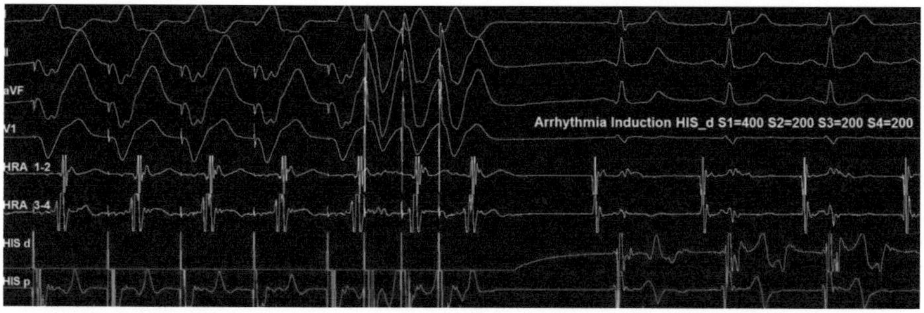

Fig. 2.9 Representation of programmed and coupled ventricular pacing to induce ventricular arrhythmias. As indicated in the text, the authors perform 8-beat drive pacing as S1 (in this case 400 ms) coupled with gradually increasing pacing until retrograde atrioventricular refractoriness occurs, by triple coupled beats as S2, S3, and S4 until refractoriness occurs (in this case S2, S3, and S4 are 200 ms). The *upper part of the figure* shows the DI, DII, aVF, and V1 surface ECG derivations. The pulse amplitude is 5 mA; the duration is 2 ms (the refractory ventricular conduction is at 2 mA with a duration of 2 ms). The *yellow color* denotes the right atrium (RA: 1–2 is distal, 3–4 is proximal); the *green color* denotes the His bundle (HIS: d is distal, p is proximal), which in this image is in the right ventricular apex. In this case the authors prefer to use quadripolar diagnostic catheters. (Image created by C. Sardu, WorkMate Claris polygraph, Abbot)

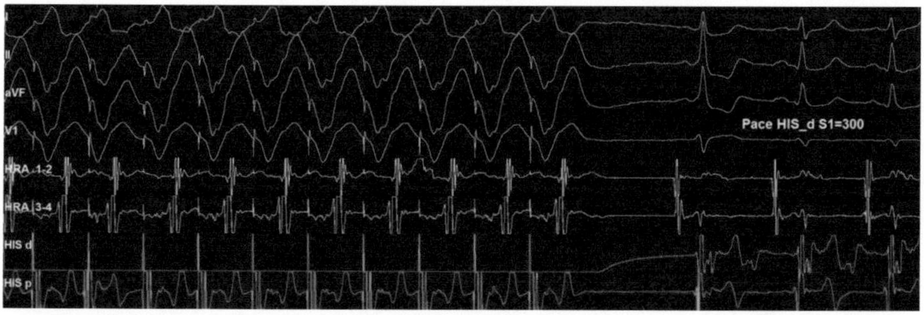

Fig. 2.10 Representation of programmed burst ventricular pacing for ventricular arrhythmia induction. As indicated in the text, the authors perform gradually increasing and continuous burst pacing from 300 to 200 ms and/or until induction of a ventricular arrhythmia. In this case the burst pacing is indicated as S1 at 300 ms. The *upper part of the figure* shows the DI, DII, aVF, and V1 surface ECG derivations. The pulse amplitude is 5 mA; the duration is 2 ms. The *yellow color* denotes the right atrium (RA: 1–2 is distal, 3–4 is proximal); the *green color* denotes the His bundle (HIS: d is distal, p is proximal) placed in the right ventricular apex. In this case the authors prefer to use quadripolar diagnostic catheters. (Image created by C. Sardu, WorkMate Claris polygraph, Abbot)

refractoriness is reached, as a conduction interval in which there is no ventricular capture with pacing [7]. Therefore, the coupled S2 interval is increased by 10–20 ms and repeated as coupled programmed stimulation introducing the secondary (S3) and then the third (S4) coupled extrastimulus in the same modality as discussed earlier in the text (Figs. 2.8 and 2.9). Similarly to atrial arrhythmia induction, during each pacing train we observe a resting period of 4 s, and we register and note every

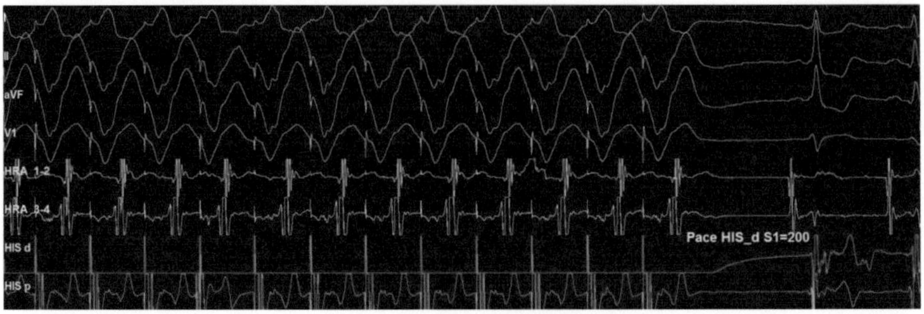

Fig. 2.11 Representation of programmed burst ventricular pacing for ventricular arrhythmia induction. As indicated in the text, the authors perform gradually increasing pacing from 300 to 200 ms and/or until induction of a ventricular arrhythmia. In this case the burst pacing is indicated as S1 at 200 ms. The *upper part of the figure* shows the DI, DII, aVF, and V1 surface ECG derivations. The pulse amplitude is 5 mA; the duration is 2 ms. The *yellow color* denotes the right atrium (RA: 1–2 is distal, 3–4 is proximal); the *green color* denotes the His bundle (HIS: d is distal, p is proximal) placed in the right ventricular apex. In this case the authors prefer to use quadripolar diagnostic catheters. (Image created by C. Sardu, WorkMate Claris polygraph, Abbot)

arrhythmic event that occurs. Therefore, we stop the S1 programmed stimulation, switching the S1 interval from 600 to 500 and/or 400 ms (F9), and repeat the same process until S4 (the third coupled extrastimulus) no longer conducts or tachycardia is triggered [2]. Completing all this programmed pacing protocol from a 600- to 400-ms S1 interval, and until refractoriness of the triple coupled extrastimulus (S2–S3–S4) occurs, we may start burst pacing to achieve a clinical ventricular arrhythmia [9] (Figs. 2.10 and 2.11). This second type of programmed pacing modality is different from coupled pacing for ventricular arrhythmias. The burst pacing is due to continuous ventricular pacing from 300 ms and down to 10–20 ms until 200 ms [9] (Figs. 2.10 and 2.11). When pacing protocols are completed through S4 by at least two S1 cycle lengths (600 and 500 and/or 400 ms), with trough burst pacing until 200 ms, we may choose another pacing site such as the right ventricular outflow tract [10]. Indeed, the programmed pacing protocol to induce ventricular arrhythmias is started over again at the second site of pacing [10]. Few authors advocate introduction of the fourth (S5) extrastimulus during ventricular tachycardia induction [7, 11]. This is a relevant discussion point, and authors express different opinions about the fourth extrastimulus to induce ventricular tachycardias [7, 11]. In fact, we have to report a great discrepancy of results regarding the specificity and sensitivity of the fourth coupled pacing protocol, to induce a ventricular arrhythmia [7, 11]. Moreover, we may speculate that it may be enough to perform a coupled pacing maneuver from the right ventricle with two different drives (500, 500, and/or 400 ms), until refractoriness of the third coupled extrastimulus (S2, S3, and S4) occurs, and at two different sites to induce ventricular tachycardias [11]. As referred to by the authors, once the process of pacing at two different ventricular sites has been completed, Isuprel (isoproterenol/isoprenaline) may be used to enhance cardiac conduction, repeating all the programmed pacing induction protocol [12]. This is not the case for patients with known or suspected coronary artery disease, in that there is no indication to use the drug [12]. Once the pacing protocols are fully performed in baseline conditions, and during Isuprel administration, and there is no arrhythmia induction, we have to consider the study result as negative. Similarly to atrial arrhythmia induction, during these pacing protocols we have to choose a standard value of the pacing pulse amplitude in milliamperes, and a duration in milliseconds. This is really important in the case of ventricular tachycardia induction, to avoid the capture of ventricular sites not so close to the tip position of the pacing catheter. Moreover, in the case of ventricular arrhythmia induction, the values of the pacing pulse amplitude and duration have to be calculated and then chosen as the lowest values to have ventricular local capture, and to avoid the induction of nonspecific ventricular arrhythmias and/or ventricular fibrillation [9]. As part of this, we have to remember that during ventricular pacing, we may also induce supraventricular arrhythmias. In this different case, we have to look at the retrograde atrioventricular conduction during ventricular pacing, and the modality of the arrhythmia induction (Fig. 2.12). In fact, in the case of supraventricular arrhythmia induction, the relationship between the ventricular signal and the atrial signal may be diagnostic for the induced tachycardia [13] (Fig. 2.12). Moreover, added to the retrograde atrioventricular activation pathway by an accurate interpretation of endocavitary electrograms, a shorter ventricular–atrial (VA) interval of <70 ms may

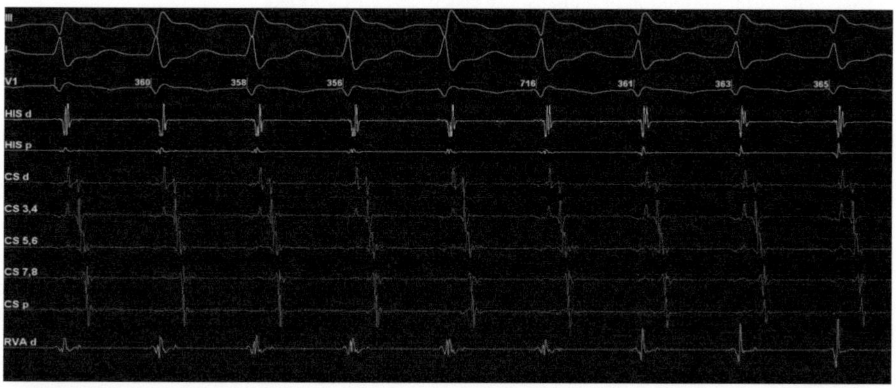

Fig. 2.12 Representation of induction of supraventricular tachycardia by programmed ventricular pacing. The induced arrhythmia has a mean cycle length of 358 ms, with eccentric retrograde atrioventricular activation and first atrial retroactivation in the proximal coronary sinus (CS). As can be seen, this tachycardia has an atrioventricular conduction of 1/1, with a retrograde ventricular–atrial (VA) interval > 70 ms. This is the case in supraventricular tachycardia induced via a concealed retrograde bypass accessory pathway (discussed in the text). The *upper part of the figure* shows the DI, DII, aVF, and V1 surface ECG derivations. The pulse amplitude is 10 mA; the duration is 2 ms. The *yellow color* denotes the His bundle (HIS: 1–2 is distal, 3–4 is proximal) placed in the right ventricular apex; the *red color* denotes the CS (d is distal, p is proximal), in this case placed in the CS ostium. In this case the authors prefer to use a quadripolar diagnostic catheter for HIS and a decapolar catheter for CS (discussed in the text). (Image created by C. Sardu, WorkMate Claris polygraph, Abbot)

indicate atrioventricular node re-entry [13]. Conversely, a more prolonged VA interval, such as a VA interval of >70 ms, points toward atrioventricular re-entry tachycardia through a bypass tract and/or atrial tachycardia [13] (Fig. 2.12). In this case ventricular pacing after the arrhythmia induction may be used to perform pacing maneuvers, such as reset and/or entrainment pacing, to differentiate the clinical diagnosis for longer VA interval tachycardias. This point will be discussed in a separate chapter.

References

1. Scheinman MM, Morady F. Invasive cardiac electrophysiologic testing: the current state of art. Circulation. 1983;67:1169–73.
2. Josephson ME. Clinical cardiac electrophysiology: techniques and interpretations. 4th ed. Philadelphia: Lippincott Williams & Williams; 2008.
3. Manolis AS, Cameron J, Deering T, Han EH, Estes NAM. Sensitivity and specificity of programmed atrial stimulation for induction of supraventricular tachycardias. Clin Cardiol. 1988;11:307–10.
4. Csanadi Z, Klein GJ, Yee R, Thakur RK, Li H. Effect of dual atrioventricular node pathways on atrioventricular reentrant tachycardia. Circulation. 1995;91:2614–8.
5. Brignole M, Menozzi C, Sartore B, Barra M, Monducci I. The use of atrial pacing to induce atrial fibrillation and flutter. Int J Cardiol. 1986;12:45–54.
6. Calvo D, Atienza L, Jalife J, Martínez-Alzamora N, Bravo L, Almendral J, et al. High-rate pacing-induced atrial fibrillation effectively reveals properties of spontaneously occurring paroxysmal atrial fibrillation in humans. Europace. 2012;14:1560–6.
7. Brugada P, Green M, Abdollah H, Wellens HJ. Significance of ventricular arrhythmias initiated by programmed ventricular stimulation: the importance of the type of ventricular arrhythmia induced and the number of premature stimuli required. Circulation. 1984;69:87–92.
8. Wellens HJJ, Brugada P, Stevenson WG. Programmed electrical stimulation of the heart in patients with life-threatening ventricular arrhythmias: what is the significance of induced arrhythmias and what is the correct stimulation protocol? Circulation. 1985;72:1–7.
9. Spielman SR, Farshidi A, Horowitz LN, Josephson ME. Ventricular fibrillation during programmed ventricular stimulation: incidence and clinical implications. Am J Cardiol. 1978;42:913–8.
10. Mann DE, Luck JC, Griffin JC, Herre JM, Limacher MC, Magro SA, et al. Induction of ventricular tachycardia using programmed stimulation: value of third and fourth extrastimuli. Am J Cardiol. 1983;52:501–6.
11. Herre JM, Mann DE, Luck JC, Magro SA, Figali S, Breen T, et al. Effect of increased current, multiple pacing sites and number of extrastimuli on induction of ventricular tachycardia. Am J Cardiol. 1986;57:102–7.
12. Sardu C, Carreras G, Katsanos S, Kamperidis V, Pace MC, Passavanti MB, et al. Metabolic syndrome is associated with a poor outcome in patients affected by outflow tract premature ventricular contractions treated by catheter ablation. BMC Cardiovasc Disord. 2014;14:176.
13. Platia EV, Reid PR. Nonsustained ventricular tachycardia during programmed ventricular stimulation: criteria for a positive test. Am J Cardiol. 1985;56:79–83.

3 How to Induce Arrhythmias with Isoprenaline

Sorin Pripon

3.1 Introduction

Isoprenaline is an isopropylamine analog of Epinephrine, which has solely a β1 and β2 effect. It acts nonselectivelly on betareceptors located in the heart, bronchi, skeletal muscle and alimentary tract. It was used in asthma for its bronchodilator effect, but at the moment there are other β2 selective agonists that are preferred. In cardiology, it is commonly used for its positive chronotropic effect.

S. Pripon (✉)
Cardiologie et Soins Intensifs, Centre Hospitalier de Bigorre, Tarbes, France

G. Cismaru (ed.), *Arrhythmia Induction in the EP Lab*,
https://doi.org/10.1007/978-3-319-92729-9_3

3.2 Mechanism of Action

Isoprenaline is a β1-adrenergic receptor agonist, causing the receptor to respond as if exposed to noradrenaline. Basically, the receptor activates the adenylate-cyclase, which then produces cAMP that stimulates the heart.

In other words, as an agonist of the G-protein-coupled receptor, isoprenaline causes a conformational change in the receptor, which in turn catalyzes the dissociation of GDP from the heterotrimeric G-protein-α subunit. The α subunit-GTP produces an activation of the adenylate -cyclase. cAMP will activate the protein kinase A (PKA). PKA causes phosphorylation of the sodium and calcium channels in the 'open' conformation and potassium channels in the 'closed' conformation. The effect on the calcium channels is depolarization of the nodal cells of the heart and of the myocardial cells. The effect on the sodium channels is depolarization of the myocardial cells, and the effect on the potassium channels is limitation of the hyperpolarization in nodal cells and myocardial cells. The final effect is increase in inotropism and chronotropism with higher heart rate and stroke volume.

3.2.1 Chemical Structure

Isoprenaline is the isopropyl analog of adrenaline which is derived from amino acids phenylalanine and tyrosine. The chemical formula of isoproterenol is $C_{11}H_{17}NO_3$ (Fig. 3.1).

Fig. 3.1 Chemical structure of isoproterenol

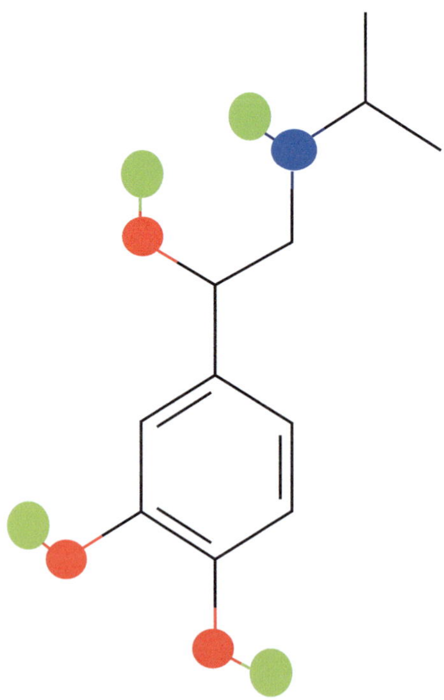

3.3 Pharmacology of Isoprenaline

Isoproterenol is rapidly absorbed when administered intravenously. Conolly found a mean half-life of 29 s (range 10–67 s) [1]. The initial inactivation of isoprenaline is mainly due to a rapid clearance from the circulation. The authors also demonstrated that a dose of 0.063 μg/kg/min given in 1 min increased the heart rate by 15 bpm with recovery of the normal heart rate after 8 min. In another patient a dose of 0.44 μg/kg/min injected during 30 min increased the heart rate by 30 bpm, which returned to normal after the end of the infusion, with a half-life of 10 min.

Isoprenaline is metabolized (and thus becomes inactive) in the liver and other tissues, by the enzyme that degrades catecholamines: catechol-O-methyltransferase.

Excretion of isoprenaline after administration is primarily renal, the main metabolite of isoprenaline being the sulfate.

3.4 Doses of Isoprenaline

Isoprenaline is classically used during electrophysiological studies, with an initial bolus of 0.02–0.06 mg, and subsequently an infusion rate of 2 to 20 micrograms/min.

In their study, Choi et al. used a bolus of 10–30 μg of isoprenaline or an infusion of 1–16 μg/min for VT induction. In case of failure to induce VT, a bolus of epinephrine 20–100 μg was injected and programmed ventricular stimulation repeated [2].

Oral et al. also used an infusion of isoprenaline for atrial fibrillation induction: infusion was started at 5 μg/kg/min and increased every 2 min to 10, 15, and 20 μg/kg/min [3] (Fig. 3.2).

In the study of Markel et al., isoprenaline was infused for VT induction in patients treated with procainamide and quinidine [10]. The dose was 0.03 μg/kg/min.

In the study of Brembilla-Perot et al., isoprenaline was started at a dose of 0.5 μg/min with a total dose of 6–20 μg [14].

In the study of Goy, isoprenaline was infused 1–6 μg/min, in order to assess the effects of propafenone in patients with WPW syndrome [26].

In the study of Hariman et al., isoprenaline 0.5 μg/min was infused to facilitate AVNRT induction [13].

In the study of Brugada et al., an initial dose of 0.25–1 μg/min was infused and doubled every 3–5 min to obtain increase in the heart rate. A mean dose of 2.2 μg was given for induction of reentrant tachycardia through an accessory pathway [15].

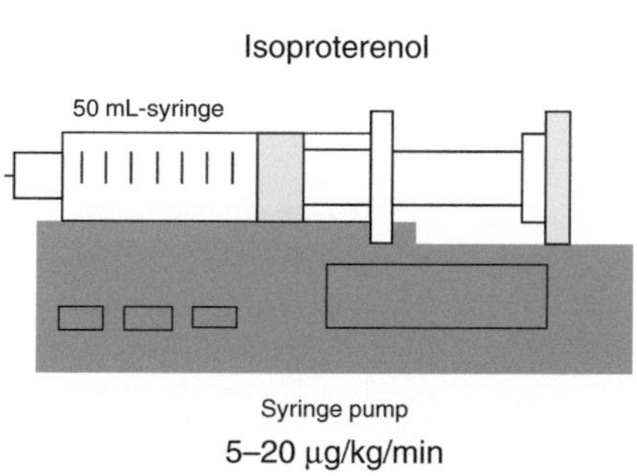

Fig. 3.2 Doses of isoprenaline

In the study of Brownstein et al., a dose of 0.5–3.0 µg/min was infused to facilitate AVNRT induction [12].

3.4.1 **Infusion Preparation** (Tables 3.1 and 3.2)

1. The patient should not eat ou drink (fasting) for at least 8 h before the electrophysiological study.
2. Patients are instructed to stop medication, five times the half-life prior to electrophysiological testing, unless instructed otherwise.
3. An IV line is started using aseptic technique.
4. An infusion pump is necessary for isoprenaline administration.
5. Isoprenaline can be diluted in distilled water, 5% dextrose or saline.
6. 1 mL Isoprenaline vial contains 0.2 mg isoproterenol. DILUTION: 5 vials of Isoprenaline (total = 5 × 0.2 = 1 mg) into a 50 mL syringe with 45 mL of 0.9% normal saline or 5% dextrose into the syringe → 1000 µg Isoprenaline in 50 mL solution. Therefore, 1 mL solution contains 20 µg of isoproterenol.
7. Give 1–2 mL solution bolus (20–40 µg of isoprenaline), and then continue with the infusion.
8. Infusion of isoproterenol is started at a dose of 5 µg/min and then increased up to the desired effect of heart rate increase.
9. Continuous monitoring of EKG, heart rate, and 1-min assessment of blood pressures and symptoms for each infusion stage.
10. Isoprenaline infusion is increased by 5 µg/min at 2-min intervals up to a maximum of 20 µg/min.

Table 3.1 Doses of Isoprenaline

Time	Dose
Start	Bolus of 20 µg Isoprenaline
2 min	5 µg/min
4 min	10 µg/min
6 min	15 µg/min
8 min	20 µg/min

Table 3.2 How to prepare the Isoprenaline infusion

Understanding the numbers:

You are ordered to give 0.1 µg/kg/min Isoprenaline to your 58-year-old, 70 kg patient for induction of orthodromic reentrant tachycardia

1. The Isoprenaline vial has 0.2 mg/mL meaning 200 µg in 1 mL
2. You should take 5 vials of Isuprel in the 50 mL syringe with 45 mL of serum (0.9% NaCl)
3. This gives a concentration of 1000 µg in 50 mL meaning 20 µg/mL
4. 0.1 x 70 kg/min = 7 µg/min should be injected, equivalent of 420 µg/h
5. The infusion rate should be set at 12 mL/h

11. An Adrenaline bolus of 200–300 μg may be used if clinical arrhythmia cannot be induced by programmed atrial or ventricular stimulation under Isoprenaline alone.
12. Intravenous Metoprolol 5 mg over 2–3 minutes, repeated up to 3 times if necessary, at 5 minutes interval, up to a maximum total dose of 15 mg can be administered in case of incessant arrhythmia under isoproterenol. Alternatively, intravenenous Atenolol can be used at a dose of 2,5–7,5 mg (in general 5 mg over 5 minutes), or Esmolol IV bolus of 0,5 mg/Kg over 1 min followed by an infusion of 3-5 mg/kg/hour.

3.5 Isoprenaline's Side Effects

Systemic vasodilatation produced by isoprenaline can induce severe hypotension. Systolic blood pressure increases and diastolic blood pressure decreases, with decrease in renal and mesenteric perfusion.

Isoproterenol increases the myocardial need for oxygen and also decreases the coronary supply by shortening the diastole. Therefore it should be avoided in case of trivascular or unstable ischemic heart disease (Tables 3.3 and 3.4).

Min Park et al. described three cases of ventricular fibrillation, after isoprenaline infusion during electrophysiological study: two of them after infusion of 10 μg/min for AF triggering spontaneous atrial foci, and the third patient after having received 3 μg/min isoprenaline for PVC induction [4].

Table 3.3 Contraindications for Isoprenaline

Contraindications for isoprenaline
• Unstable Angina/Acute coronary syndrome (except for cases with high-grade A-V block with severe poorly-tolerated bradycardia); Extreme caution is needed in multi-vessel ischemic heart disease
• Tachycardia > 130 BPM
• Significant ventricular or atrial hyperexcitability (e.g. non-sustained/sustained bursts of ventricular tachycardia or repetitive atrial tachyarrhythmias)
• Digitalis toxicity
• Known isoprenaline hypersensitivity/allergy or intolerance .... Hypovolemic shock with circulatory collapse.... Concomitent use of halogenated inhalational anesthetics
• Caution in hyperthyroidism
• Caution in elderly individuals

Table 3.4 Side effects of Isoproterenol

Side effects of isoproterenol
• Headache
• Atrial or ventricular tachyarhythmias; Angina, myocardial ischemia
• Hypotension
• Flushing
• Restlessness, Tremor (Shaking)
• Weakness
• Dizziness... nausea

3.6 Animal Studies

Hertring found that in rats, isoprenaline has a urinary excretion in about 8 h [5].

In dogs, an isoprenaline dose of 0.27–0.64 μg/kg increased the heart rate by 22–80 bpm, with a half-life of 0.5–2.5 min after the end of the infusion. The excretion was 46% urinary and 0,5% biliary [6].

Balazs et al. injected 1 mg/kg isoprenaline in dogs, induced ventricular tachycardia with associated myocardial necrosis in histological specimens but without increase in cardiac serum enzymes [7]. Contrary to these findings, Osborne et al. reported elevations of CPK and LDH after 1 mg/kg isoprenaline injection [8].

Gopinath et al. injected 2 mg/kg isoprenaline in dogs and induced ventricular tachycardia and fibrillation with increased serum enzymes of myocardial necrosis [9].

3.7 Human Studies

Infusion of isoprenaline during sinus rhythm at a perfusion rate of 0.03 μg/kg/min leads to decrease in the sinus cycle from 792 to 568 ms (approximately 25%), decrease of the QT interval from 386 to 346 ms, decrease of the paced QRS complex from 185 to 182 ms, decrease of the ventricular effective refractory period (VRP) from 238 to 208 ms and functional VRP from 261 to 227 ms, and decrease of the induced ventricular tachycardia cycle length from 311 to 291 ms [10].

In several studies it was demonstrated that isoprenaline modifies the refractory period of the atrioventricular node [11]. Brownstein et al. [12] also showed that it modifies the refractory period of both the slow and fast atrio-ventricular node pathways. In individuals who practice sport and have an enhanced vagal tonus, isoprenaline lowers the refractory period of the node thus making it possible to induce atrioventricular node reentrant tachycardia. Hariman confirmed the mechanism and also showed that intranodal reentry is induced because of an enhanced retrograde ventriculoatrial conduction under isoprenaline [13]. In patients with effort-induced arrhythmias, isoprenaline is a very good agent for junctional supraventricular tachycardia inducibility (sensibility of 92% and specificity of 100%) and also for ventricular tachycardia inducibility (sensibility of 75% and specificity of 95%) [14].

Brugada et al. induced atrioventricular node reentrant tachycardia in 21 patients, and they explained the mechanism by enhanced anterograde conduction through the AV node [15].

Regarding the induction of atrial fibrillation in the electrophysiology laboratory, in the study of Oral et al. [3] different doses of isoprenaline were used. Initially, a dose of 5 μg/min was started, and then the dose was increased to 10.15 and 20 μg/min at 2-min intervals, in 20 patients without history of atrial fibrillation and in 80 patients with a history of atrial fibrillation. In the control group, only 1 of 20 patients had inducible AF. In the group with paroxysmal atrial fibrillation, AF was inducible in 67 of 80 patients (84%). The overall sensitivity was 88% and the specificity 95%. Only in four patients (5%) isoprenaline was stopped due to secondary effects. The mechanism for atrial fibrillation induction has multiple explanations: decrease in the sinus node cycle length, decrease of the refractory period of the pulmonary veins and left atrium, and early atrial depolarizations. Furthermore, ganglionic plexus stimulation and increase in the parasympathetic tone can initiate and perpetuate atrial fibrillation by a decrease of the effective atrial refractory period and induction of triggered activity with spontaneous depolarizations.

3.7.1 Reversal of Encainide by Isoprenaline Infusion in Patients with Ventricular Tachycardia

Encainide is a drug of the IC class of antiarrhythmics. It acts by blocking sodium channels. It affects the antegrade and retrograde conduction through the slow nodal pathway, fast nodal pathway, and accessory pathways. This is why encainide is effective in patients with AVNRT and reentrant tachycardias through accessory pathways. The most important effect in patients with AVNRT is the block of the retrograde pathway [16, 17].

Akhtar et al. [18] tested the effects of encainide on 16 patients with AVNRT and 16 patients with WPW syndrome before and after administration of isoprenaline. Encainide (75–200 mg) was effective in protecting against reentrant arrhythmias in 14 of the 16 patients from the AVNRT groups and 8 of the 16 patients from the orthodromic reentrant tachycardia (ORT) group.

After isoprenaline infusion, it reversed the effects of encainide in 10 patients from the AVNRT group and 2 patients from the ORT group; the reentrant arrhythmia in both groups became reinducible. In conclusion isoprenaline reversed the effects of encainide both on the A-V node and accessory pathways, both the antegrade and retrograde conductions [18, 19].

The authors showed that drugs able to control the arrhythmia in the basal state, could fail to control it during special physiological or psychological conditions. Administration of isoprenaline could mimic these changes in autonomic tone, similar to the changes that occur during psychological and physiological stress. Patients that remained non-inducible after isoprenaline infusion had a better prognosis without recurrences of tachycardia during follow-up.

3.7.2 Reversal of Flecainide by Isoprenaline Infusion in Patients with Accessory Pathways

Flecainide administration in patients with WPW syndrome affects the accessory pathway and also the atria, ventricles, and AV node. Flecainide blocks both antegrade and retrograde conduction through the accessory pathway preventing the development of orthodromic reentrant tachycardia (ORT). It also has a protective role against atrial fibrillation at the level of the atria by prolonging the atrial refractory period. Isoproterenol was demonstrated to block the effects of flecainide and other class IC antiarrhythmic drugs: propafenone and encainide. The study of Manolis et al. demonstrates that isoprenaline blocks the effects of flecainide both on the accessory pathway and atrial tissue [20].

In malignant forms of WPW syndrome, flecainide is used to prevent the 1:1 conduction to the ventricles, preventing very high ventricular rates and hemodynamic instability. In the study of Brembilla-Perot et al. [21], isoprenaline was administrated in patients with WPW syndrome treated with flecainide. Isoprenaline reverses the effects of flecainide and allows 1:1 conduction through the accessory pathway from the atrium to the ventricles. For those patients, catheter ablation was proposed because in special adrenergic conditions like physiological or psychological stress, the effects of flecainide on the accessory pathway properties might be reversed, with 1:1 conduction to the ventricles.

Helmy et al. [22] verified if the isoprenaline test can predict recurrences of ORT in patients treated with flecainide. In 17 patients treated with flecainide, they injected isoprenaline and reinduced ORT in 8 patients. They explained their results by the reversal effects of isoprenaline on patients treated with flecainide. Flecainide prolongs the retrograde refractory period of the accessory pathway, prolongs the AH and HV interval, and reduces conduction through the AV node both retrograde and antegrade. After isoprenaline infusion, the retrograde refractory period of the accessory pathway was shorter, the AH and HV intervals decreased, and the conduction through the AV node was faster both in antegrade and retrograde directions. Those patients that presented ORT after isoprenaline infusion also presented arrhythmia recurrences at follow-up.

3.7.3 Reversal of Quinidine by Isoprenaline Infusion in Patients with Ventricular Tachycardia

In the study of Markel et al. [10], isoprenaline was administered in patients with ventricular tachycardia treated with quinidine. Thirteen patients were studied at baseline and after quinidine treatment. In patients treated with quinidine, isoprenaline shortened the cycle length during sinus rhythm, the QT interval during pacing, and the ventricular refractory period. Isoprenaline also reduced the tachycardia cycle length during VT and the QRS duration during pacing in patients treated with quinidine.

3.7.4 Reversal of Procainamide by Isoprenaline Infusion in Patients with Ventricular Tachycardia

Procainamide is a class IA antiarrhythmic drug that acts on sodium channels with a blocking effect. It is used for the treatment of ventricular premature beats and ventricular tachycardia. In 18 patients treated with procainamide, isoprenaline shortened the cycle length during sinus rhythm, the cycle length during ventricular tachycardia, the QT interval during pacing, the QRS duration during pacing, and the ventricular refractory period. After isoprenaline infusion, the number of extrastimuli needed to induce ventricular tachycardia was reduced in 13 patients [23].

3.7.5 Reversal of Propafenone with Isoprenaline Infusion in Patients with Accessory Pathways

Propafenone is a good antiarrhythmic drug for patients with paroxysmal supraventricular tachycardias (SVT). The response to digoxin, beta-blockers, or calcium-channel blockers is not ideal, with frequent recurrences under these medications. Propafenone has a better efficacy profile than the three above-mentioned drugs in patients with paroxysmal SVT [24, 25].

Goy et al. [26] studied the reversal of antiarrhythmic effects of propafenone by isoprenaline infusion. They infused isoprenaline in ten patients treated with propafenone for WPW syndrome. Propafenone prolonged the antegrade and retrograde refractory period of the accessory pathway, but isoprenaline reversed the effects of propafenone with reduction of the refractory periods of the accessory pathway. The sinus cycle length decreased after isoprenaline, and the tachycardia cycle length also decreased as isoprenaline accelerates conduction through both the accessory pathway and the AV node. Patients that were uninducible after propafenone treatment received isoprenaline infusion that facilitated induction of reentrant tachycardia. On the other hand, the stress test during treatment with propafenone did not produce modifications similar to those that appeared during isoprenaline infusion. Therefore isoprenaline can predict recurrences of tachycardia after propafenone treatment in patients with WPW, but the stress test fails to do so.

References

1. Conolly ME, Davies DS, Dollery CT, George CF. Resistance to B-adrenoceptor stimulants (a possible explanation for the rise in asthma deaths). Br J Pharmacol. 1971;43:389–402.
2. Choi EK, Kumar S, Nagashima K, Lin KY, Barbhaiya CR, Chinitz JS, et al. Better outcome of ablation for sustained outflow-tract ventricular tachycardia when tachycardia is inducible. Europace. 2015;17:1571–9.
3. Oral H, Crawford T, Frederick M, Gadeela N, Wimmer A, Dey S, et al. Inducibility of paroxysmal atrial fibrillation by isoproterenol and its relation to the mode of onset of atrial fibrillation. J Cardiovasc Electrophysiol. 2008;19:466–70.
4. Min Park Y, Soo Lee H, Lim RA. Inadvertently developed ventricular fibrillation during Electrophysiologic study and catheter ablation: incidence, cause, and prognosis. Korean Circ J. 2013;43:474–80.
5. Hertring G. The fate of 3H-iso-proterenol in the rat. Biochem Pharmacol. 1964;13:1119–28.
6. Conolly ME, Davies DS, Dollery CT, Morgant CD, Patersons JW, Sandler M. Metabolism of isoprenaline in dog and man. Br J Pharmacol. 1972;46:458–72.
7. Balazs T, Ohtake S, Cummings JR, Noble JF. Ventricular extrasystoles induced by epinephrine, nicotine, ethanol and vasopressin in dogs with myocardial lesions. Toxicol Appl Pharmacol. 1969;15:189.
8. Osborne BE. Uses and applications of electrocardiography in toxicology. Proc Eur Soc Study Drug Toxic. 1974;15:85.
9. Gopinath C, Thuring J, Zayed I. Isoprenaline-induced myocardial necrosis in dogs. Br J Exp Pathol. 1978;59:148–57.
10. Markel ML, Mile WM, Luck JC, Klein LS, Prytowsky EN. Differential effect of isoproterenol on sustained ventricular tachycardia before and during procainamide and quinidine antiarrhythmic drug therapy. Circulation. 1993;87:783–92.

11. Coelho A, Palileo E, Ashley W, Swiryn S, Petropoulos T, Welch WJ, et al. Tachyarrhythmias in young athletes. J Am Coll Cardiol. 1986;7:237–43.
12. Brownstein SL, Hopson RC, Martins JB, Aschoff AM, Olhansky B, Constantin L, et al. Usefulness of isoproterenol in facilitating atrioventricular nodal reentry tachycardia during electrophysiological testing. Am J Cardiol. 1988;61:1037–41.
13. Hariman RJ, Gomes JAC, El-Sherif N. Cathecolamine-dependent atrioventricular nodal reentrant tachycardia. Circulation. 1983;67:681–6.
14. Brembilla-Perrot B, Terrier de la Chaise A, Pichène M, Pernot C, Cherrier F. Diagnostic value of the isoproterenol test in effort tachycardia. Arch Mal Coeur Vaiss. 1986;79(3):302–13.
15. Brugada P, Facchini M, Wellens HJJ. Effects of isoproterenol and amiodarone and the role of excercise in initiation of circus movement tachycardia in the accessory atrio-ventricular pathway. Am J Cardiol. 1986;57:146–9.
16. Niazi I, Nacarelli G, Dougherty A, Rinkenberger R, Tchou P, Akhtar M. Treatment of atrioventricular node reentrant tachycardia with encainide: reversal of drug with isoproterenol. J Am Coll Cardiol. 1989;13:904–10.
17. Niazi I, Naccarelli G, Dougherty A, Rinkenberger R, Tchou P, Akhtar M. Treatment of AV nodal reentrant tachycardia with encainide: reversal of drug effect on the AV node with isoproterenol. J Am Coll Cardiol. 1989;13:904–10.
18. Akhtar M, Niazi I, Naccarelli G, Tchou P, Rinkenberger R, Dougherty AH, et al. Role of adrenergic stimulation by isoproterenol in the reversal of effects of encainide in supraventricular tachycardia. Am J Cardiol. 1988;62:45L–9L.
19. Naccarelli G, Berns E, Dougherty A, Jenkins M, Rinkenberger R. Isoproterenol reversal of encainide effect in the Wolf-Parkinson-White syndrome. PACE. 1987;10:95–100.
20. Manolis AS, Estes NAM. Reversal of electrophysiology effects of flecainide on accessory pathway by isoproterenol in the Wolff-Parkinson-White syndrome. J Am Coll Cardiol. 1989;64:194–8.
21. Brembilla-Perot B, Admant P, Le Helloco A, Pernot C. Loss of efficacy of flecainide in the Wolff Pakinson-White syndrome after isoproterenol administration. Eur Heart J. 1985;6:1074–8.
22. Helmy I, Scheinman MM, Sharkey H, Herre JM, Griffin JC. Isoproterenol reversal of flecainide effects in patients with accessory pathways. Circulation. 1987;76(suppl IV):IV–69.
23. Wu D, Wyndham C, Amat-Y-Leon F, Denes P, Dhingra RC, Rosen K. The effects of ouabain on induction of atrio-ventricular nodal reentrant paroxysmal tachycardia. Circulation. 1978;57:1171–9.
24. Bauernfeind RA, Wyndham CR, Dhingra RC, et al. Serial electrophysiologic testing of multiple drugs in patients with atrioventricular nodal reentrant paroxysmal tachycardia. Circulation. 1980;62:1341–9.
25. Wellens HJ, Duren DR, Liem KL, Lie KI. Effect of digitalis in patients with paroxysmal atrial ventricular nodal tachycardia. Circulation. 1975;52:77–88.
26. Goy JJ, Fromer M, Schlaepfer J, Kappenberger L. Electrophysiologic effects of intravenous propafenone at rest, during isoproterenol infusion and during exercise in the Wolff-Parkinson-White syndrome. Am J Cardiol. 1991;68:681–4.

4 How to Induce Arrhythmias with Orciprenaline

Nihal El Kamar and Lazhare Naouar

4.1 Introduction

Orciprenaline is a bronchodilator used for the treatment of asthma and COPD. The other name for orciprenaline is metaproterenol. It has a β2 selective effect with no effect on alpha receptors. It stimulates the smooth muscle in the bronchiole, uterus and arterioles supplying the skeletal muscle [1].

Orciprenaline, which is an orcinol isomer of adrenaline (Fig. 4.1), was developed primarily because of the disadvantages of isoprenaline. Isoprenaline is unstable, is short acting and has some cardiac side effects unlike orciprenaline.

N. El Kamar · L. Naouar (✉)
Augusta-Krankenhaus, Klinik fur Kardiologie, Dusseldorf, Germany

G. Cismaru (ed.), *Arrhythmia Induction in the EP Lab*,
https://doi.org/10.1007/978-3-319-92729-9_4

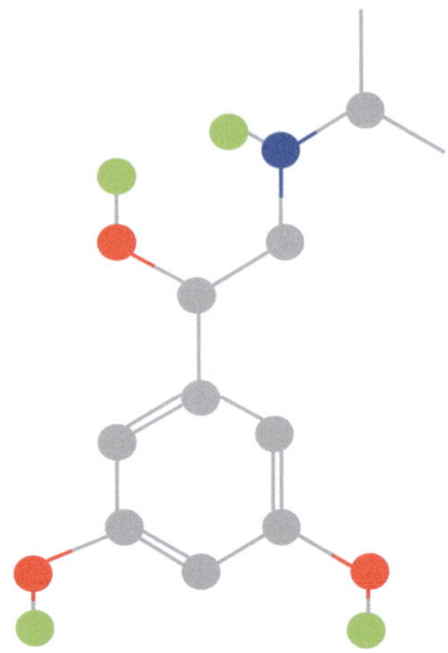

Fig. 4.1 Chemical structure of orciprenaline

4.2 Pharmacology

Orciprenaline is metabolised in the liver and eliminated in the urine. The onset of action is very fast: 15 sec after intravenous administration, 15 min after oral administration and 60 sec after inhalation.

After intravenous infusion, less than 15% of the drug is absorbed intact [1].

The duration of action is approximately 30–45 min after intravenous administration and 1–4 h after oral administration.

The half-life of orciprenaline is 6 h.

In mice the intravenous lethal dose (LD50) is 114 mg/kg [2].

4.3 Mechanism of Action

Orciprenaline stimulates the β-receptors of the enzyme adenylyl cyclase that permits conversion of ATP to cAMP. Increased levels of cAMP induce relaxation of the bronchioles and the smooth muscles from arterioles and the uterus.

Orciprenaline is a more specific β2 agonist than isoproterenol hence causes less tachycardia and hypotension at equipotent doses.

4.4 Doses of Orciprenaline

Intravenous orciprenaline was used in the 1960s to increase the heart rate of patients with AV block or sinus node disease. Redwood (1968) reported the effect of orciprenaline in 23 patients with AV block. For the preparation of infusion, the authors used 5 mg orciprenaline in 30 mL of saline (165 μg/mL). The infusion rate was 150–200 μg/min. The rate was raised every 2–3 min to increase the heart rate or to obtain a concentration of 200–300 μg/min of orciprenaline (Figs. 4.2 and 4.3).

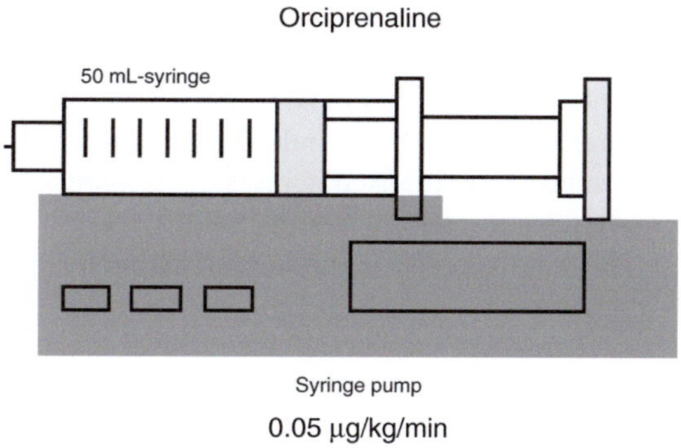

Fig. 4.2 Doses of orciprenaline

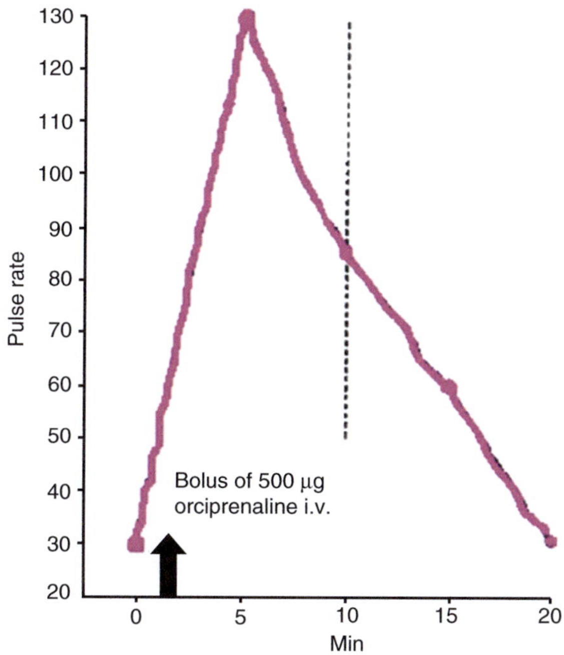

Fig. 4.3 A bolus of 500 μg of orciprenaline produces an increase in the heart rate from 30 to 50 bpm in 1 min, with a peak at 130 bpm after 5 min. Modified after Hock et al. [7]

Hock et al. [3] compared the effects of orciprenaline bolus and intravenous infusion. Bolus was given in a dose of 250–1000 μg and the intravenous infusion of a dose of 7–14 μg/min. The ventricular rate increased from 30 to 50 bpm after 1 min and with a maximum of 130 bpm after 5 min of 500 μg of bolus. This effect lasted for 20 min. Boluses should be repeated within a time interval of 15–20 min in case of a longer mapping procedure.

During a continuous infusion of orciprenaline, a dose of 7 μg/min increased the heart rate from 30 to 50 bpm and maintained it for a longer period of time.

Parwani et al. [4] used a cumulative dose of 0.5 mg orciprenaline over 15 min in a patient with AVNRT ablation.

Strohmer et al. [5] used an infused dose of 0.5 μg/min of orciprenaline to increase the heart rate by 30% of the basal heart rate in patients with non-inducible AVNRT.

Sinkovec et al. [6] and Abd Al-Hassan et al. [7] used a dose of 0.25–0.5 mg of orciprenaline for AVNRT induction, respectively, in a study of 104 and 40 patients. Weismuller et al. [8] used boluses of orciprenaline between 0.25 and 1 mg to induce arrhythmias in patients with AVNRT.

The way of preparing an orciprenaline infusion is described in Fig. 4.4, Tables 4.1 and 4.2.

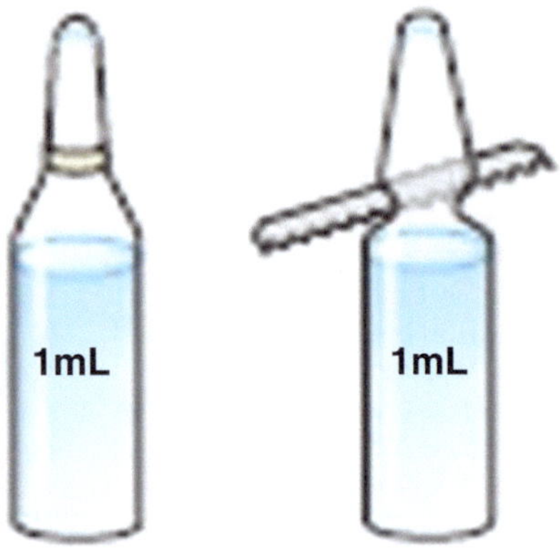

Orciprenaline vial 0.5 mg/1mL

Fig. 4.4 Orciprenaline vial contains 0.5 mg in 1 mL solution. It is produced in Germany and mostly used for electrophysiology studies

Table 4.1 How to prepare the Orciprenaline infusion

Understanding the numbers:

You are ordered to give 0.05 μg/kg/min Orciprenaline to your 58-year-old, 70 kg patient for induction of orthodromic reentrant tachycardia

1. The Orciprenaline vial has 0.5 mg/mL meaning 500 μg in 1 mL
2. You should take 1 vial of Orciprenaline in the 50 mL syringe with 49 mL of serum (0.9% NaCl)
3. This gives a concentration of 500 μg in 50 mL meaning 10 μg/mL
4. 0.05 x 70 kg/min = 3.5 μg/min should be injected, equivalent of 210 μg/h
5. The infusion rate should be set at 21 mL/h

Table 4.2 Preparation of the patient for Orciprenaline infusion

1.	The patient should not eat for at least 8 h before the electrophysiological study
2.	Patients are instructed to stop medication five times the half time prior to electrophysiological testing unless instructed otherwise
3.	Orciprenaline can be given either as a bolus or as an infusion
4.	In case of boluses, the dose is 0.25 mg that can be repeated until the heart rate increases by 50%. Usual dose is 0.5–1 mg
5.	In case of infusion, the dose is 0.5 μg/kg/min to increase the heart rate by 50%

4.5 Preparation

- The patient should not eat for at least 8 h before the electrophysiological study.
- Patients are instructed to stop medication for a period of time equal to 5 times the medication's plasma half-life.
- Orciprenaline can be given as a bolus or as an infusion.
- When boli are used, a 0.25 mg dose should be repeated until a 50% heart rate increase is obtained. Usual dose is 0.5–1 mg.
- In case of infusion, the recommended dose of 0.5 μg/kg/min until a 50% heart rate increase is obtained.

4.6 Side Effects of Orciprenaline

Major side effects are uncommon, with an incidence of more than 10% for:

- Tremor
- Nervousness

Minor side effects have an incidence of less than 10%:

- Diaphoresis
- Headache
- Pharyngitis
- Dizziness
- Insomnia
- Weakness
- Nausea
- Exacerbation of asthma
- Hypertension (Tables 4.3 and 4.4)

Table 4.3 Side effects of Orciprenaline

Side effects of Orciprenaline
• Pallor
• Drowsiness
• Headache
• Nervousness
• Tremor
• Weakness
• Nausea, vomiting, epigastric distress
• Anxiety, restlessness
• Paradoxical bronchoconstriction
• Fear, irritability, trembling
• Dizziness, lightheadedness
• Sweating
• Hypokalemia (at high doses)

Table 4.4 Contraindications for Orciprenaline

Contraindications for Orciprenaline
• Known sensitivity to the drug or other sympathomimetic amines
• Hypertrophic obstructive cardiomyopathy
• Pheocromocytoma
• Thyrotoxicosis
• Unstable ischemic heart disease
• Recent myocardial infarction

4.7 Electrophysiological Effects of Orciprenaline

4.7.1 Animal Studies

When administered to dogs in bronchodilator doses, orciprenaline has a minimal effect on blood pressure and heart rate, compared to isoproterenol. In dogs it induces a decrease in the diastolic blood pressure with an increase in the systolic blood pressure due to an elevation of the heart rate. In isolated heart preparations, orciprenaline demonstrated an increase in inotropic and chronotropic effects. In the heart preparations where the pacemaker tissue is conserved, orciprenaline acts on the sinus node but with an affinity that is lower than the one of isoproterenol [3]. Surprisingly, intravenous administration of orciprenaline to guinea pigs failed to induce arrhythmias. Furthermore, adrenaline administration after orciprenaline infusion failed to induce arrhythmias, despite high doses of 3–30 times greater than the one usually causing arrhythmias [9].

Komadina et al. [10] studied the electrophysiological effects of orciprenaline on five mongrel dogs and found a decrease in the atrial and ventricular refractory period, AH interval and Wenckebach cycle length. Orciprenaline also increases the conduction through the AV node and His-Purkinje. Heart rate increased from 126 to 220 bpm, and MAP decreased from 118 to 92 mmHg.

The same authors associated aminophylline with orciprenaline and found a significant decrease of the AH and HV interval, atrial refractory period, Wenckebach cycle length and ventricular refractory period compared to the only use of aminophylline but no further decrease compared to orciprenaline alone. The heart rate increases with the combination and on the other hand the MAP decreased [11].

Kirchhof et al. [12] studied catecholaminergic ventricular tachycardia in mice expressing triadin (TRD) and junction (JCN). They used isolated hearts with continuous infusion of orciprenaline (1.7 μM) and demonstrated spontaneous repetitive episodes of VT in TRD × JCN hearts.

4.7.2 Human Studies

Gattiker et al. [13] studied the hemodynamic effects of orciprenaline with a dose of 4 and 8 μg/min and compared them to dopamine and epinephrine. They found that in patients early after cardiac surgery, orciprenaline has a positive chronotropic effect and should be restricted to patients with low heart rate. In case that a high dose of orciprenaline 8 μg/min was needed, the combination of orciprenaline 4 μg + dopamine was superior to orciprenaline 8 μg/min alone to increase the heart rate.

4.7.2.1 Orciprenaline for Supraventricular Extrabeats Induction Due to 1:2 Atrioventricular Conduction

Dual atrioventricular pathway can cause AVNRT and rarely 1:2 conduction of a sinus beat to ventricles, leading to supraventricular extrabeats. The treatment in this particular form of PAC is a modulation of the slow pathway. Pott et al. [14] reported 5 patients with dual AV node physiology with 1:2 conduction after orciprenaline infusion. The catheter modulation of the slow pathway was safe in those five patients with a good long-term outcome.

4.7.2.2 Orciprenaline for AVNRT Induction

Weber et al. [15] reported laser catheter coagulation in 10 patients with AVNRT. In all patients tachycardia was induced after orciprenaline administration. The dose used in this study was 1.5–2 μg/kg infusion. Ablation was effective in all 10 patients.

Sinkovec et al. [6] injected orciprenaline in 104 patients for AVNRT induction before and after catheter ablation. The dose used by the authors was a bolus of 0.25–0.5 mg with an addition of 0.5–1 mg atropine in case of non-induced AVNRT before ablation. Catheter ablation was effective in 96% after 16 months of follow-up.

Weismuller et al. [8] studied 131 patients with AVNRT. In 100 patients, tachycardia was inducible in the basal state, while orciprenaline was needed for tachycardia induction in 31 patients. They demonstrated that orciprenaline should be given after ablation only in patients who require orciprenaline for tachycardia induction before ablation. The dose used was 0.25–1 mg of orciprenaline adjusted to obtain a heart rate of 50% above the basal heart rate.

Heinroth et al. [16] used orciprenaline 0.5–1 mg for arrhythmia induction in a study of 78 patients with AVNRT.

Strohmer et al. [5] infused 0.5 μg/min orciprenaline with the aim to increase the heart rate above 30% of the basal heart rate in patients with AVNRT. They discovered that 40% of the patients with AVNRT also presented other associated arrhythmias: atrial fibrillation in 19%, focal atrial tachycardia in 8%, atrial flutter in 6% and AV reentrant tachycardia in 4%. More than 2 types of tachycardia were present in 12% of patients.

Reents et al. [17] followed 49 patients with cryoablation for AVNRT and identified recurrences in 11 patients (22.4%). Orciprenaline was used for AVNRT induction in 8 of 11 (73%) patients representing the recurrence group and 21 of the 38 (55%) patients representing the non-recurrence group.

4.7.2.3 Orciprenaline for PVC Induction

Orciprenaline can be used for PVC induction before and after catheter ablation. Even in patients with very frequent PVCs (>30,000/24 h), these can be absent during the electrophysiology study. Catheter ablation of PVCs is mainly based on activation mapping. Wojdyla-Hordynska et al. [18] performed catheter ablation in

109 patients with and without concomitant structural heart disease. Orciprenaline was used for PVC induction before and after ablation. The success rate of ablation was 85.3%.

Ventricular tachycardia originating in the left ventricular outflow tract can also be induced by ventricular stimulation after orciprenaline infusion. Frey et al. [19] reported three cases of VT from LVOT, two with tachycardia induced in the basal state and one necessitating orciprenaline for induction. In all three patients, catheter ablation abolished VT.

4.7.2.4 Orciprenaline for AF Induction

After a successful pulmonary vein isolation, recurrences are mainly due to reconnection of the pulmonary veins. Orciprenaline can unmask residual PV connection. Brunelli et al. [20] infused orciprenaline in doses of 5–30 μg/min, and PV reconnection was verified at doses of 20–30 μg/min. Orciprenaline triggered a reconnection in 5% of the veins (3 out of 60).

The same authors added adenosine to orciprenaline to unmask PV reconnections. Compared to orciprenaline alone (which had a success rate of 5% for PV reconnection), adding adenosine further increased the efficacy of drug challenge. The combination unmasked residual connection in further 9 patients, corresponding to 88% of cases (53 out of 60). Adenosine alone unmasked PV reconnection in 73% of cases (44 out of 60).

4.7.2.5 Orciprenaline for VT Induction

Strohmer et al. [5] used orciprenaline infusion for ventricular tachycardia ablation in patient with frequent episodes of VT, treated with antiarrhythmic drugs and implanted with an ICD. Authors targeted isolated diastolic potentials and terminated VT within 5 s. Afterwards, they enlarged the initial lesions by RF applications. After ablation, VT was non-inducible by programmed ventricular stimulation both before and after adrenaline infusion.

References

1. Hatch F, McKellop K, Hansen G, et al. Relative bioavailability of metaproterenol in humans utilizing a single dose stable isotope approach. J Pharm Sci. 1986;9:886–90.
2. Rona G, Chappel C, Zsoter T, et al. Myocardial lesions circulatory and electrocardiographic changes produced by isoproterenol in the dog. Rev Can Biol. 1959;18:83–94.
3. Hock C, Toh CCS, Khoo TH. Clinical trial of orciprenaline in bradyarrhythmias. Singapore Med J. 1975;16:100–5.
4. Parwani AS, Schroder A, Blaschke D, et al. Third- degree AV block sensitive to prednisolone 72 hour post ablation ov AVNRT. Clin Case Rep. 2017;5:671–4.
5. Strohmer B, Hwang C. Ablation of postinfarction ventricular tachycardia guided by isolated diastolic potentials. Europace. 2003;5:375–80.
6. Sinkovec M, Vrecar I, Sustar M, Pernat A, Rakovec P, et al. Electrophysiological and clinical characteristics of atrioventricular nodal reentrant tachycardia and long-term success of radiofrequency catheter ablation. Zdravniški vestnik. 2010;79:280–9.

7. Abd Al-Hassan A. Radiofrequency ablation of typical atrioventricular node reentry tachycardia in dual and multiple AV nodal pathways. Med J Babylon. 2015;3:745–54.
8. Weismuller P, Kuly S, Brandts B, et al. Is electrical stimulation during administration of catecholamines required for evaluation of success after ablation of atrioventricular node reentrant tachycardia? J Am Coll Cardiol. 2002;39:691–4.
9. Shanks RG, Brick I, Hutchinson K, Roddie IC. Stimulation of adrenergic beta-receptors by orciprenaline. Br Med J. 1967;1:610.
10. Rona G, Chapel C, Balazs T. An infarct-like myocardial lesion and other toxic manifestations produced by isoproterenol in the dog. Rev Can Biol. 1959;18:83.
11. Komadina KH, Carlson TA, Strollo PJ, et al. Electrophysiologic study of the effects of aminophylline and metaproterenol on canine myocardium. Chest. 1992;100:232–8.
12. Kirchhof P, Klimas J, Fabritz L, et al. Stress and high heart rate provoke ventricular tachycardia in mice expressing triadin. J Mol Cell Cardiol. 2007;42:962–71.
13. Gattiker R, Schmid E. Hemodynamic effects of dopamine, epinephrine and orciprenaline (Alupent) in patients early after cardiac surgery. Intensive Care Med. 1978;4:55–61.
14. Pott C, Wegner F, Bogeholz N, Fischer A, et al. Supraventricular extrasystole due to 1:2 atrioventricular conduction: procedural characteristics and long term follow-up of slow pathway modulation. Eur Heart J. 2013;34(suppl 1_1):P2350.
15. Weber HP, Kaltenbrunner W, Heinze A, et al. Laser catheter coagulation of atrial myocardium for ablation of atrioventricular node reentrant tachycardia. Eur Heart J. 1997;18:487–95.
16. Heinroth KM, Kattenbeck K, Sabenow I. Multiple AV nodal pathways in patients with AV nodal reentrant tachycardia-more common than expected ? Europace. 2002;4:375–82.
17. Reents T, Springer B, Hessling G, et al. Long-term follow-up after cryoablation for adolescent atrioventricular nodal reentrant tachycardia: recurrence is not predictable. Europace. 2012;14:1629–33.
18. Wojdyla-Hordynska A, Kowalsi O, Hordynski G, et al. The effect of radiofrequency catheter ablation of frequent premature ventricular complexes and arrhythmia burden on left ventricular function. Kardiol Pol. 2017;75:698–704.
19. Frey B, Kreiner G, Fitsch S, Veit S, Gossinger HD. Successful treatment of idiopathic left ventricular outflow tract tachycardia by catheter ablation or minimally invasive surgical cryoablation. PACE. 2000;23:870–6.
20. Brunelli M, Raffa S, Grobe A. Residual condution after pulmonary vein isolation with a circular multielectrode radiofrequency catheter ablation: the role of adenosine and orciprenaline during a prolonged observation time. Int J Cardiol. 2013;168:4122–31.

How to Induce Arrhythmias with Adrenaline

5

Bogdan Chendran

5.1 Introduction

Adrenaline, also known as epinephrine, is a hormone produced by the adrenal gland that was first isolated in 1895 by the Polish physiologist Napoleon Cybulski [1].

In Europe and Great Britain, the term adrenaline is preferred, but in the United States, epinephrine is used over adrenaline. However, drugs that mimic epinephrine

B. Chendran (✉)
Medicine Physique et Readaptation, Centre Hospitalier Bagnères de Bigorre, Bagnères de Bigorre, France
e-mail: bogdan.chendran@ch-bagneres.fr

G. Cismaru (ed.), *Arrhythmia Induction in the EP Lab*,
https://doi.org/10.1007/978-3-319-92729-9_5

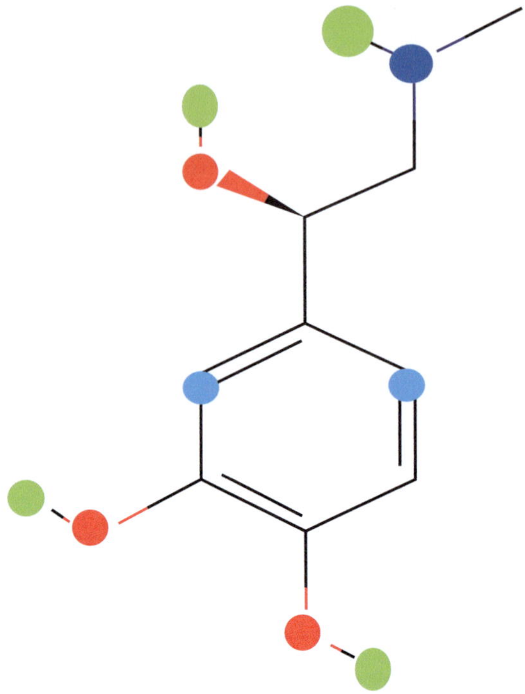

Fig. 5.1 Chemical structure: Adrenaline is a catecholamine, a sympathomimetic monoamine derived from the amino acids phenylalanine and tyrosine. It is produced both in some neurons and in the adrenal medulla. The chemical formula of adrenaline is $C_9H_{13}NO_3$ [6]

are called adrenergic, and receptors for epinephrine are also called adrenergic receptors [2].

Over 90% of the circulating adrenaline is secreted by the adrenal medulla, but only 7% of circulating noradrenaline is released by the adrenal glands.

5.1.1 Chemical Structure

Adrenaline is a catecholamine, a sympathomimetic monoamine derived from the amino acids phenylalanine and tyrosine. It is produced both in some neurons and in the adrenal medulla. The chemical formula of adrenaline is $C_9H_{13}NO_3$ [6] (Fig. 5.1).

5.2 Pharmacology of Adrenaline

In normal individuals, endogenous plasma concentrations of adrenaline are less than 10 ng/L. This concentration increases tenfold during exercise and 50-fold during physiological or emotional stresses [3].

After adrenaline is given intravenously, a significant increase in plasma concentrations appears within 1 min [4]. In cardiac reanimation, a bolus of adrenaline can produce increased plasma concentrations as 10,000–100,000 ng/L [5].

Elimination half-life of epinephrine is 1 min, with a biological half-life of 2 min.

The excretion of adrenaline is made by urine.

5.3 Mechanism of Action

Adrenaline stimulates the α1, α2, β1, β2, and β3 receptors of the sympathetic nervous system. During exercise [7] adrenaline concentration increases by secretion from the adrenal gland and produces increased blood flow to the muscles, increased cardiac output, dilatation of the pupil, and increased blood glucose [8]. It was demonstrated that adrenaline is released in response to stress [9], and a greater concentration is correlated with a state of negative feelings [10], fear, and other negative emotions [11]. It has been also demonstrated that adrenaline can enhance long-term memory and memory consolidation.

The cardiovascular effects of adrenaline are due to its activity on α- and β1-adrenergic receptors, with an only moderate activity on the β2 receptors (Tables 5.1, 5.2, 5.3 and 5.4). The inotropic and chronotropic action is due to a direct effect on the myocardium [12], with vasoconstriction in the vascular bed, leading to precapillary resistance in the skin vessels, mucosa, kidney, and also vasoconstriction at the level of the veins [13].

In terms of cardiac stimulation, adrenaline has a greater inotropic effect due to beta stimulation effect than noradrenaline.

Table 5.1 Effects of adrenaline on organs

Effects of adrenaline on organs and tissues in the body	
Organ	Effect—receptor type
Heart	Increase heart rate—β1 increase contractility—β1
Blood vessels	Vasoconstriction—α1 vasodilatation—β2
Lungs	Bronchodilation—β2
Uterus	Relaxation—β2

Table 5.2 Types of alpha receptors

Types of α-adrenergic receptor	
Receptor—site of action	Effects
α1—smooth muscle, heart, and liver	Vasoconstriction, intestinal relaxation, uterine contraction, and pupillary dilatation
α2—platelets, vascular smooth muscle, nerve termini, and pancreatic islets	Platelet aggregation, vasoconstriction, and inhibition of NE release and of insulin secretion

Table 5.3 Types of beta receptors

Types of β-adrenergic receptor	
Receptor—site of action	Effects
β1—Heart	Tachycardia
β2—Lungs	Bronchodilation smooth muscle relaxation sphincter constriction
—Gastrointestinal tract liver, uterus	
—Vascular smooth cells skeletal muscle	
β2—Fat cells	

Table 5.4 Effects of adrenaline and noradrenaline on receptors

Epinephrine and norepinephrine actions on adrenergic hormone receptors	
Receptor	Agonists
Alpha 1	NE > E
Alpha 2	E > NE
Beta 1	E = NE
Beta 2	E >> NE

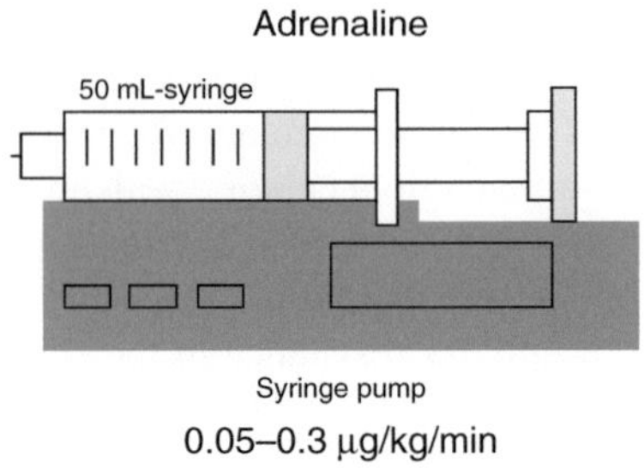

Fig. 5.2 Adrenaline dosage

Table 5.5 Dosage for Adrenaline over time unit

Time	Dose
Start	0.05 µg/kg/min
5 min	0.1 µg/kg/min
10 min	0.15 µg/kg/min
15 min	0.20 µg/kg/min
20 min	0.25 µg/kg/min
25 min	0.30 µg/kg/min

In terms of constriction of blood vessels, adrenaline has a lower effect than noradrenaline with a lesser effect on peripheral resistance and increase in the blood pressure.

5.4 Doses of Adrenaline

Adrenaline is started at a dose of 0.05 µg/kg/min and increased every 5 min by 0.05 µg/min in order to obtain increase in the heart rate of more than 50% of the basal rate or to a sinus cycle length of 600–400 ms. The medium dose used for PSVT or VT induction is 0.1–0.3 µg/kg/min [14]. After adrenaline infusion, programmed atrial and ventricular stimulation is continued to obtain the desired arrhythmia (Fig. 5.2, Tables 5.5 and 5.6).

5.5 Contraindications to Adrenaline

The most important contraindications are listed in Table 5.7.

Table 5.6 Infusion preparation for Adrenaline

Infusion preparation	
1.	The patient should not eat for at least 3 h before the electrophysiological study
2.	Patients are instructed to stop medication five times the half time prior to electrophysiological testing unless instructed otherwise
3.	An IV is started using aseptic technique
4.	An infusion pump is necessary for adrenaline administration. No bolus is given
5.	Add 3 vials of adrenaline 1 mg/mL into 50 cc syringe with 48 mL of NaCl 0.9% into the syringe
	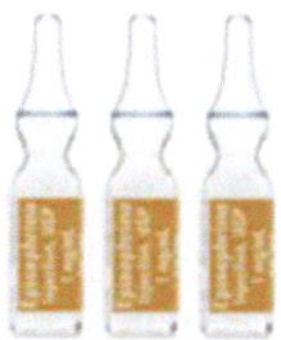
6.	Adrenaline is started at a dose of 0.05 μg/kg/min
7.	Continuous monitoring of EKG, heart rate, and 1-min assessment of blood pressures and symptoms for each infusion stage
8.	Adrenaline infusion is increased by 0.05 μg/kg/min at 5-min intervals up to a maximum of 0.3 μg/kg/min
9.	Atropine 1–3 mg may be used if clinical arrhythmia cannot be induced by programmed atrial or ventricular stimulation

Table 5.7 Contraindications for Adrenaline

Contraindications for adrenaline during electrophysiological study
• Recent (<1 week) myocardial infarction
• Unstable angina
• Hemodynamically significant left ventricular (LV) outflow tract obstruction
• Severe aortic stenosis
• Uncontrolled hypertension (blood pressure > 200/110 mmHg)
• Patients with aortic dissection or large aortic aneurysm
• Unwillingness to give consent

5.6 Side Effects of Adrenaline

The most relevant side effects of adrenaline are listed in Table 5.8.

5.7 Animal Studies

Zink et al. demonstrated that a dose of 0.5–3 μg/kg adrenaline infused in dogs induced ventricular premature contractions, bigeminism, and multifocal contractions. Ventricular fibrillation did not appear after adrenaline infusion [15].

Lessard et al. studied different doses of adrenaline that can induce ventricular arrhythmias in dogs. They found that 5 μg/kg/min can cause premature ventricular

Table 5.8 Side effects of Adrenaline

Side effects of adrenaline
• Tremor
• Headache
• Back pain
• Atrial premature contractions
• Ventricular premature contractions
• Non-sustained atrial tachycardia
• Myocardial ischemia

contractions. The arrhythmogenic effect of adrenaline can be inhibited by the addition of difluorodichlormethane [16].

In the study of Ajioka et al., adrenaline 10 μg/kg was injected in 30 dogs to verify the induction of ventricular tachycardia. Seventeen dogs had low serum potassium, and 13 dogs had normal potassium at baseline. Ventricular tachycardia appeared in 53% of the hypokalemia group and 23% of the normal potassium group. Ventricular tachycardia emerged in the first seconds after adrenaline injection. Dogs with higher percentage of arrhythmia ratio had a higher concentration of mitochondrial calcium. Calcium blockers may have a protective effect against ventricular arrhythmias in patients with hypokalemia [17, 18].

5.8 Human Studies

Adrenaline can be used for both supraventricular and ventricular arrhythmia induction. The effect on PSVT induction is close to that of isoprenaline [19, 20]. In the study of Cismaru et al., adrenaline had a good sensitivity of 82% and an excellent specificity of 100% for the induction of paroxysmal supraventricular tachycardia in patients with documented arrhythmia (Table 5.9). The authors compared 66 patients with documented PSVT with 30 patients without PSVT and obtained the sensitivity and specificity of adrenaline. These results suggest that adrenaline can be used as a substitute for isoprenaline to induce PSVT in the electrophysiology lab when the latter is missing due to manufacturing problems or difficulties in importing it from other countries [21].

Choi et al. used adrenaline boluses for outflow tract arrhythmia induction. Inability to induce ventricular premature beats or ventricular tachycardia from right or left outflow tract is a major factor that contributes to failure of the ablation procedure. The authors used either isoproterenol bolus or adrenaline bolus to induce the clinical arrhythmia. Whereas isoprenaline produces hypotension [22], adrenaline increases the blood pressure, so a different mechanism of action differentiates the two drugs. Of 25 patients with non-inducible arrhythmia in the basal state, isoprenaline was injected in 19 patients. And adrenaline was given on top of isoprenaline in six patients. The dose of adrenaline was 20–100 μg administered in bolus [23].

Antiarrhythmic drugs have the effect of inhibiting the clinical arrhythmia of the patient, and this can be proven in the electrophysiology laboratory by the inability to induce tachycardia during programmed atrial or ventricular stimulation. However, despite antiarrhythmic drugs, there are dynamic factors that change the autonomic

Table 5.9 Sensitivity and specificity of adrenaline for PSVT induction

Adrenaline's accuracy for PSVT induction			
Sensitivity	Specificity	Positive predictive value	Negative predictive value
82%	100%	100%	71%

tone and can precipitate episodes of rapid arrhythmia. Therefore, it is useful and necessary to verify induction of arrhythmia in the EP lab using adrenaline, noradrenaline, atropine, or isoprenaline, in order to check the efficacy of antiarrhythmic drugs under different physiological and pathological stresses.

5.8.1 Inhibition of Verapamil Effects by Adrenaline Infusion in Patients with PSVT

The electrophysiologic effects of adrenaline are opposite to those of verapamil, an antiarrhythmic drug used to treat patients with PSVT, the dose being 240–480 mg daily. Adrenaline was infused at a dose of 25–50 ng/kg/min, a dose that leads to plasma concentrations similar to those occurring during various physiological or pathological stresses. For the 25 ng/kg/min dose, a plasma concentration of 862 ± 226 pg/mL was obtained. These concentrations also occur during a submaximal exercise test, smoking, public speaking, mild hypoglycemia, dental extraction, or surgery. For the 50 ng/kg/min dose, a plasma concentration of 1.374 ± 477 pg/mL was obtained. The plasma concentration is similar to those obtained during the maximal exercise test, myocardial infarction, diabetic ketoacidosis, or severe hypoglycemia.

Since 10 min of infusion is required to achieve a steady-state plasma level for adrenaline, it was infused for 14 min to obtain the desired electrophysiological changes. Note that adrenaline alone injected in patients did not trigger any of the PSVT, but stimulation was performed to induce PSVT under adrenaline. Adrenaline decreased the sinus cycle from 761 to 615 ms, the AH interval, the Wenckebach point, and the refractory periods of the atrium and the ventricle. It does not change the HV interval which remains constant (46 ms) before and after adrenaline administration. This is important finding because one can test the effect of adrenaline for VT induction in the electrophysiology lab.

The conclusion of the article was that adrenaline antagonized partially or totally the effects of verapamil. This explains why some patients treated with verapamil still exhibit PSVT under physiological or pathological conditions that lead to increase concentrations of catecholamines.

5.8.2 Inhibition of Quinidine Effects by Adrenaline Infusion in Patients with VT

In the study of Hugh Calkins [24], adrenaline administration was similar to Professor Morady's protocol at Ann Arbor. Thus, adrenaline in six patients in the absence of quinidine resulted in decreased baseline sinus rhythm and decreased effective and absolute ventricular refractory period but had no effect on QRS duration. Instead, the ventricular tachycardia cycle length decreased by 30–80 ms. VT induction required two to three ventricular extrastimuli in the basal state but only one to two extrastimuli after infusion. When quinidine was administered, the basal cycle increased and the effective and functional ventricular refractory periods also increased as well as QRS duration. After adrenaline infusion, the basal cycle and refractory periods decreased but remained higher than in the basal state without quinidine. The QRS duration remained unchanged after administration infusion, i.e., increased to 194 ms compared to 196 ms after quinidine compared to 170 ms in the basal state. This is explained by the effect of adrenaline on β-adrenergic receptors and the cellular effects shortening of the myocardial action potential.

On the other hand, adrenaline has no effect on the intraventricular conduction, so it does not alter it when administered alone. Adrenaline also has no effect on intraventricular conduction when it is affected (slowed down) by quinidine. Finally the effects of quinidine are not totally antagonized by adrenaline, and the action on intraventricular conduction remains unaffected.

Nevertheless, the study of Hugh Calkins shows that infusion of adrenaline in quinidine-treated VT patients led to a decrease in ventricular refractory period compared to pre-quinidine treatment. Thus, adrenaline caused a complete antagonism of quinidine, the effects being measured by return to the initial ventricular effective refractory period.

5.8.3 Inhibition of Amiodarone Effects by Adrenaline Infusion in Patients with VT

Hugh Calkins et al. have studied the effects of adrenaline in quinidine-treated VT patients and amiodarone-treated VT patients. The protocol for the electrophysiological study with adrenaline was similar to that published in other papers of the team from Ann Arbor Michigan. Of 29 patients treated with quinidine, VT was inducible in all. After amiodarone, 5 patients of 29 became non-inducible. In four of the five patients, adrenaline was infused, but none became inducible at programmed ventricular stimulation. The calculated sensitivity of adrenaline infusion is 0% for VT induction. However the small number of patients constitutes an important limitation of the study. It should be noted that electrophysiological study was performed after 10 days of amiodarone, and the daily dose was 1800 mg. At that time the plasma concentration of amiodarone was 2 ± 0.43 mg/L.

5.8.4 Inhibition of Quinidine Effects by Adrenaline Infusion in Patients with WPW Syndrome

Isoprenaline is not endogenously produced; therefore authors studied the effects of adrenaline infusion in WPW patients treated with quinidine. To demonstrate that adrenaline can antagonize the effects of quinidine, Morady et al. administered

Table 5.10 How to prepare an Adrenaline infusion

Understanding the numbers:

You are ordered to give 0.1 μg/kg/min of Adrenaline to your 58 –year-old, 70 kg patient to facilitate ventricular tachycardia induction by programmed stimulation

1. The Adrenaline vial has 1mg/1 ml.
2. You should take 3 vials of Adrenaline in the 50 ml syringe with 47 ml of serum (0,9% NaCl)
3. This gives a concentration of 3 mg/50 ml meaning 0.06 mg/ml.
4. 0.1μgx70kg=7μg should be injected, thus 0.007 mg equivalent of 0.117 ml/min
5. We will start the perfusion rate: 0.117x60= 7ml/hour.

quinidine in 18 patients with orthodromic reentrant tachycardia (ORT), and 9 of them adrenaline was infused afterward. As a result, eight patients became non-inducible after quinidine, but adrenaline reversed the effects of quinidine, so five patients became reinducible. Thus, the study demonstrates that an increased level of adrenaline similar to increased concentrations occurring during various physiological or pathological stresses can reverse the effects of quinidine in patients with WPW and ORT.

The way of preparing an adrenaline infusion is described in Table 5.10.

References

1. Szablewski L. Glucose homeostasis and insulin resistance. Sharjah: Betham Science; 2011. p. 68.
2. Dorner J. Comparative experimental studies on circulatory effect of adrenaline and epinephrine. Arch Kreislaufforsch. 1954;21:88–115.
3. Clutter WE, Cryer PE. Plasma dose—response studies with noradrenaline and adrenaline in man. Prog Biochem Pharmacol. 1980;17:84–9.
4. Mazkereth R, PAret G, Ezra D, Aviner S, Peleg E, Rosenthal T, et al. Epinephrine blood concentrations after peripheral bronchial versus endotracheal administration of epinephrine in dogs. Crit Care Med. 1992;20:1582–7.
5. Raymondos K, Panning B, Leuwer M, Brechelt G, Korte T, Niehaus M, et al. Absorption and hemodynamic effects of airway administration of adrenaline in patients with severe cardiac disease. Ann Intern Med. 2000;132:800–3.
6. Engelman K, Portnoy B, Portnoy B, Lovenberg W. A sensitive and specific double-isotope derivative method for the determination of catecholamines in biological specimens. Am J Med Sci. 1968;255:259–68.
7. Goldstein DS, Dionne R, Sweet J, Gracely R, Brewer B, Gregg R, et al. Circulatory, plasma-catecholamine, cortisol, lipid, and psychological responses to a real-life stress (third molar extractions)—effects of diazepam sedation and of inclusion of epinephrine with the local-anesthetic. Psychosom Med. 1982;44:259–72.
8. Bell DR. Medical physiology: principles for clinical medicine. 3rd ed. Philadelphia: Lippincott Williams and Wilkins; 2009. p. 312.
9. Christensen NJ, Videbaek J. Plasma catecholamines and carbohydrate metabolism in patients with acute myocardial infarction. J Clin Invest. 1974;54:278–86.
10. Clutter W, Bier DM, Shah SD, Cryer PE. Epinephrine plasma metabolic clearance rates and physiologic thresholds for metabolic and hemodynamic actions in man. J Clin Invest. 1980;66:94–101.

11. Toth M, Ziegler M, Sun P, Gresak J, Risbrough V. Impaired conditioned fear response and startle reactivity in epinephrine-deficient mice. Behav Pharmacol. 2013;24:1–9.
12. Morady F, Nelson SD, Kou WH, Pratley R, Schmaltz S, De Buitleir M, et al. Electrophysiologic effects of epinephrine in humans. J Am Coll Cardiol. 1988;11:1235–44.
13. Christensen NJ, Alberti KG, Alberti KG, Brandsborg O. Plasma catecholamines and blood substrate concentrations: studies in insulin induced hypoglycemia and after adrenaline infusions. Eur J Clin Investig. 1975;5:415–23.
14. Cohen G, Holland B, Holland B, Sha J, Goldenberg M. Plasma concentrations of epinephrine and norepinephrine during intravenous infusions in man. J Clin Invest. 1959;38:1935–41.
15. Zink J, Sasyniuk BI, Dresel PE. Halothane-epinephrine induced cardiac arrhythmias and the role of the heart rate. Anesthesiology. 1975;43:548–55.
16. Lessard Y, Desbrousses S, Paulet G. Cardiac arrhythmia in dogs under the action of adrenaline and difluorodichloromethane. C R Seances Soc Biol Fil. 1977;171:1270–82.
17. Ajioka M, Sugiyama S, Ogawa K, Satake T, Ozawa T. Mechanism of cardiac arrhythmias induceb by epinephrine in dogs with hypokalemia. J Am Coll Cardiol. 1986;8:1373–9.
18. Darbar D, Smith M, Smith M, Morike K, Morike K, Roden DM. Epinephrine-induced changes in serum potassium and cardiac repolarization and effects of pretreatment with propranolol and diltiazem. Am J Cardiol. 1996;77:1351–5.
19. Brembilla-Perrot B, Terrier de la Chaise A, Pichene M, Aliot E, Cherrier F, Pernot C. Isoprenaline as an aid to the induction of catecholamine dependent supraventricular tachycardias during programmed stimulation. Br Heart J. 1989;61:348–55.
20. Stellbrink C, Diem B, Schauerte P, Brehmer K, Schuett H, Hanrath P. Differential effects of atropine preexcitation by vagal stimulation or isoproterenol and isoproterenol on inducibility of atrioventricular nodal reentrant tachycardia. J Interv Card Electrophysiol. 2001;5:463–9.
21. Cismaru G, Rosu R, Muresan L, Puiu M, Andronache M, Hengan E, et al. The value of adrenaline in the induction of supraventricular tachycardia in the EP lab. Europace. 2014;16:1634–8.
22. Przybylski J, Chiale PA, Halpern MS, Nau GJ, Elizari MV, Rosenbaum MB. Unmasking of ventricular administration. Circulation. 1980;61:1030–7.
23. Choi EK, Kumar S, Nagashima K, Lin KY, Barbhaiya CR, Chinitz JS, et al. Better outcome of ablation for sustained outflow-tract ventricular tachycardia when tachycardia is inducible. Europace. 2015;17:1571–9.
24. Calkins H, Sousa J, Atassi R. Reversal of antiarrhythmic drug effects by epinephrine: quinidine versus amiodarone. J Am Coll Cardiol. 1992;19:347–52.

How to Induce Arrhythmias with Atropine

6

Calin Siliste and Roxana-Nicoleta Siliste

6.1 Introduction

Atropine sulfate is the sulfate salt of atropine (Fig. 6.1), which is an alkaloid found in its natural form together with scopolamine and hyoscyamine in the plant *Atropa belladonna.* The first synthesis of atropine was achieved by Richard Willstätter, a German organic chemist whose studies in the field of alkaloids brought him the 1915 Nobel Prize for Chemistry.

C. Siliste (✉)
Cardiology Department, University Emergency Hospital, University of Medicine and Pharmacy "Carol Davila", Bucharest, Romania

R.-N. Siliste
Internal Medicine and Cardiology Department, Coltea Clinical Hospital, University of Medicine and Pharmacy "Carol Davila", Bucharest, Romania

G. Cismaru (ed.), *Arrhythmia Induction in the EP Lab*,
https://doi.org/10.1007/978-3-319-92729-9_6

Fig. 6.1 Chemical structure of atropine

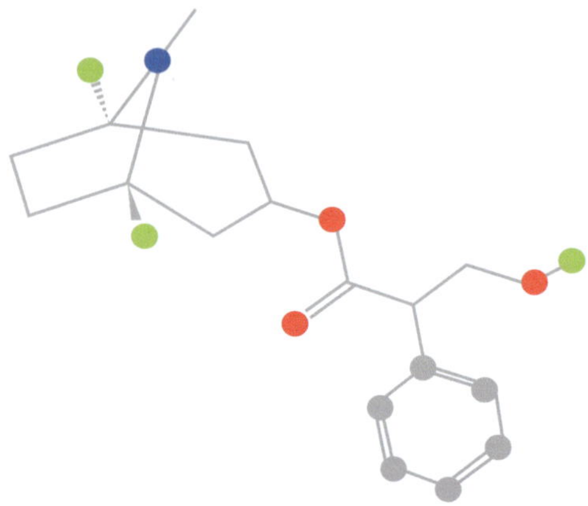

Atropine is now synthetically produced and widely used as an antidote for organophosphate poisoning but also for other therapeutic purposes, due to its extracardiac (antispasmodic, mydriatic, anticholinergic) as well as cardiac effects. The primary cardiac actions of atropine are linked to its vagolytic effect and consequently vary depending on sensitivity of conducting system to autonomic stimuli. The main effects of atropine are an increase in the automaticity of the sinus node (SN) and facilitation of atrioventricular (AV) nodal conduction capability, thus increasing heart rate and enhancing AV conduction. Beside these effects, there are other types of action and consequences on the heart, linked either to increased heart rate or to unopposed action of the sympathetic nervous system [1].

The positive chronotropic and dromotropic effects of atropine have been recognized since the beginning of the twentieth century, when Robinson and Draper [2] noticed that atropine abolished atrial premature beats and when Wilson [3] demonstrated its ability to induce AV dissociation. These findings paved the way for the research on cardiac effects of the drug, initially oriented to AV conduction but also encouraged the widespread use of atropine as anti-bradycardia agent even in the setting of myocardial infarction [4–7].

6.2 Mechanism of Action

Atropine is an anticholinergic or parasympatholytic agent, which antagonizes the effects of the parasympathetic system, mediated by acetylcholine (ACh) as primary neurotransmitter. ACh binds to muscarinic receptors (M), of which are of several types, the most dominant in humans being M2 receptor. It has been stated that all clinically significant effects of the parasympathetic system are mediated by changes in ion channel activity that occurs in response to activation of M2 muscarinic cholinergic receptors, following release of the acetylcholine [8] (Figs. 6.2 and 6.3).

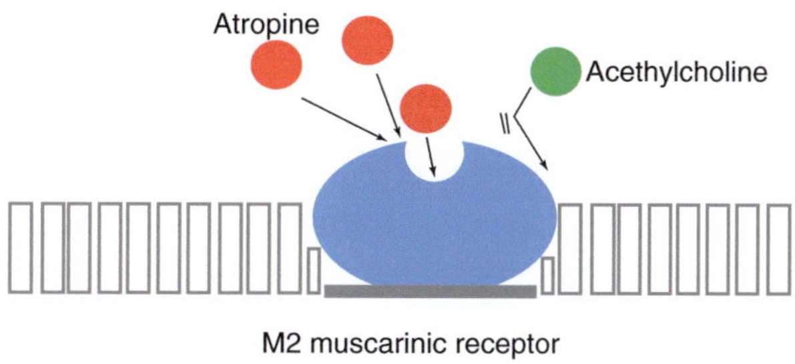

Fig. 6.2 M2 muscarinic receptor for atropine

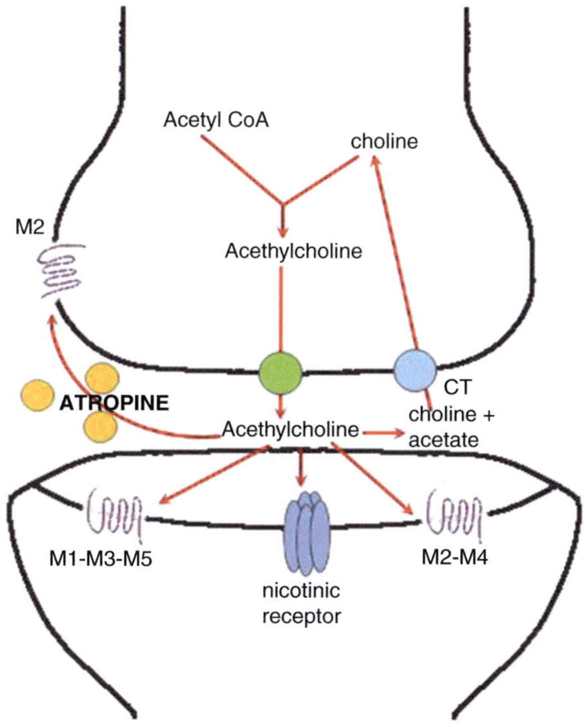

Fig. 6.3 Muscarinic M1 to M5 and nicotinic receptors influenced by acetylcholine and atropine

The effect of coupling of ACh to the M2 receptor induces change in the cardiac ion channels function by two mechanisms. The first involves direct G protein- dependent regulation of ion channel activity, and the second one involves indirect regulation of ion channel activity through modulation of cAMP-dependent responses. The direct effect is manifested through regulation of G protein-coupled inward rectifying K channels, expressed mainly in atrial, SN, and AV node cells. Activation of this pathway results in parasympathetic inhibition (slowing) of heart rate and has contributory effects in slowing impulse conduction through the AV node. The effect of the activation of M2 muscarinic receptor on cAMP is more complex, demonstrating both inhibitory and stimulatory responses [8].

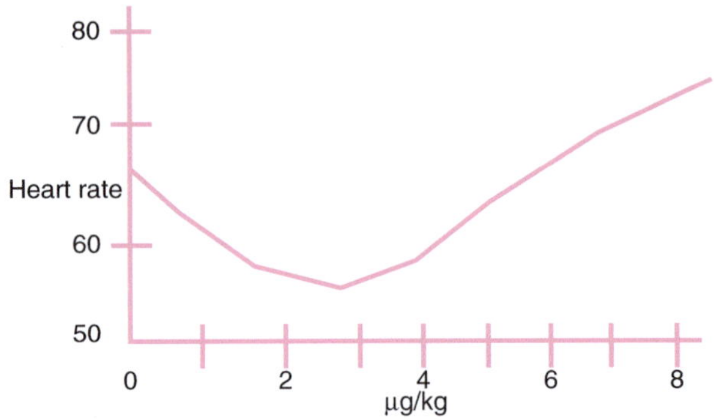

Fig. 6.4 Biphasic effect of atropine depending on administered dose. At low doses it can induce bradycardia, and in high doses it induces tachycardia

It is notable that at low doses (<0.5 mg), atropine can exert a paradoxical effect, linked to a vagotonic stimulation of the central nervous system, with sinus bradycardia and decrease of normal respiratory sinus arrhythmia. At higher doses this effect is masked by muscarinic blockade at the cardiac level [9] (Fig. 6.4).

Usual doses of atropine (>0.5 mg) abolish various types of vagal reflex-mediated bradycardia or asystole and also prevent or abolish the negative chronotropic effect produced by other parasympathomimetic drugs. Atropine also ameliorates the AV conduction when an incomplete block is noted. In some patients with complete heart block, the resultant escape rhythm rate (junctional origin) may be accelerated by atropine.

The effect of intravenous atropine infusion on maximum heart rate is nonlinear, depending on the amount of the drug in the peripheral compartment, and demonstrates further delays in onset of action by approximately 7–8 min after drug administration [10].

6.3 Pharmacokinetics

After intravenous administration, atropine disappears quickly from the bloodstream, and it is subsequently distributed in all tissues. The pharmacokinetics of atropine is nonlinear after an intravenous bolus administration of 0.5–4 mg [11]. Atropine's plasma protein binding is about 44% and saturable in the 2–20 μg/mL concentration range. Atropine is capable of crossing the placental barrier and entering the fetal circulation but is not found in amniotic fluid. Much of the drug is metabolized by enzymatic hydrolysis, particularly in the liver only 13–50% being excreted unchanged in the urine. Metabolism, however, has been noted to be inhibited by organophosphates. Traces may still be found in various secretions, including milk

[12]. The major metabolites of atropine are noratropine, atropine-n-oxide, tropine, and tropic acid [13].

6.4 Doses of Atropine

Recommended dosage for adult patients is 0.5–1 mg, with repeat doses given after 5 minutes [14] (Fig. 6.5). The use of atropine in pediatric patients is not well studied, but the initial dose is 0.01–0.03 mg/kg [12] (Table 6.1).

6.5 Preparation

1. The patient should not eat for at least 8 h before the electrophysiologic study.
2. Patients are instructed to stop medications five times the half time prior to electrophysiological testing, unless instructed otherwise.
3. Atropine is given as a bolus of 1 mg followed by supplementary boluses of 1 mg until the heart rate increases more than 50% of the basal rate.

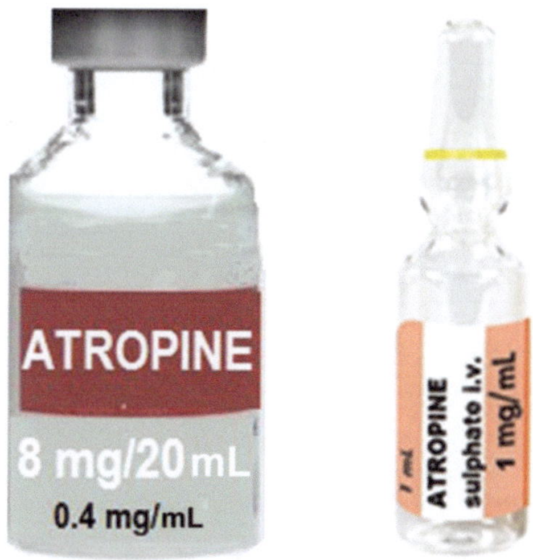

Fig. 6.5 Atropine sulfate injection is a sterile isotonic solution of atropine sulfate in water for injection

Table 6.1 Atropine administration during EP study

Time	Dose
Start	1 mg bolus
After 5 min	1 mg bolus
After 5 min	1 mg bolus
Do not exceed 3 mg atropine in total	

6.6 Contraindications to Atropine

Contraindications to atropine are summarized in Table 6.2.

6.7 Side Effects of Atropine

Most of the side effects of atropine are directly related to its antimuscarinic action and are summarized in the Table 6.3.

6.8 Electrophysiologic Effects of Atropine

The two main cardiac effects of atropine are linked to its vagolytic action on the sinus and the AV node and are noted to increase the heart rate and to facilitate the AV conduction, respectively. However, this action varies according to the sensitivity of various structures and sometimes can lead to unpredictable and apparently paradoxical effects [15, 16].

Table 6.2 Contraindications for Atropine

Contraindications and precautions for atropine
• *Coronary artery disease*
– Restrict the total dose to 2–3 mg (maximum 0.03–0.04 mg/kg) in order to reduce the detrimental effects on increasing myocardial oxygen consumption
• *Glaucoma*
– May precipitate acute glaucoma
• *Prostate hypertrophy*
– May precipitate complete urine retention
• *Pyloric obstruction*
– May transform partial organic pyloric obstruction in complete obstruction

Table 6.3 Side effects of Atropine

Side effects of atropine
• Dryness of the mouth
• Blurred vision
• Photophobia
• Anhidrosis (can produce secondary heat intolerance)
• Constipation (more often in elderly)
• Difficulty in micturition (more often in elderly)
• Hypersensitivity reactions (skin rashes that can progress to exfoliation)

6.8.1 Effect on the Sinus Node

6.8.1.1 Sinus Node Automaticity

The normal response of the SN to the usual dose of atropine (0.5 mg or more) is an acceleration of the sinus rate. This acceleration is progressive or, less frequently, may have a biphasic behavior, characterized by an initial slowing of the heart rate which is followed by a positive chronotropic effect within 2–3 min [1]. In patients with suspected SN disease, it is mandatory to assess the response to changes in autonomic tone. This can be done by using drugs that act on the vegetative nervous system-like atropine, isoproterenol, and propranolol. Atropine is the most used agent to assess parasympathetic tone in this setting [15].

There exists no commonly accepted definition for the normal cardiac rhythm response to usual doses of atropine; however, it is generally accepted that the increase in heart rate must be greater than 90 bpm or in the range of 20–50% over the control rate, with most patients with intrinsic SN dysfunction demonstrating a reduced response below these limits [15–17]. Evidence for this, are rather weak, with a number of arguments present, but still with a lot of gaps in evidence, such as the lack of standardized pharmacologic testing and studies, a poorly establishing dosing range, questionable or uncertain published sensitivity and specificity values in studies, and the observation that normal response to atropine testing does not definitively exclude sinus node disease. However, the noted presence of a failure to increase the patient's heart rate after atropine administration may be a marker of sinus node dysfunction, requiring workup.

Another measure of SN automaticity is sinus node recovery time (SNRT), defined as the duration of the pause after a standardized period of atrial pacing. Excessive overdrive suppression is a parameter that can also differentiate between intrinsic and extrinsic SN dysfunction. Atropine in usual dosages lowers the SNRT without eliminating the overdrive suppression of the SN but with abolition of the oscillations in sinus cycle length following rapid atrial pacing [15]. Of note, in some patients, atropine may paradoxically lengthen the SNRT by a poorly understood mechanism, probably linked to its effects on the conduction in the SN [18].

The study by Morton and Thomas demonstrated that a dose of 0.4 mg of atropine decreases the heart rate and doses above 0.6 mg increase the heart rate. Approximately 1.2–2 mg is necessary for complete vagal blockade [19].

6.8.1.2 Sinoatrial Conduction

Sinoatrial conduction time (SACT) is another parameter used for the diagnosis of SND. It can be measured indirectly, after resetting the SN, in the absence of overdrive inhibition, with atrial stimulation and taking into account the assumption that anterograde and retrograde conduction between the SN and the pacing sites are identical. The effect of atropine on this parameter has been studied extensively, with normal subjects demonstrating shortened SACT, while patients with intrinsic SND failing to manifest such shortening [20].

6.8.1.3 Sinus Node Denervation

Atropine is commonly used in combination with propranolol for the so-called pharmacologic autonomic denervation. The heart rate (HR) after this blockade is considered to be the "intrinsic heart rate" (IHR) representing the cardiac frequency in the absence of any autonomic nervous system influence. IHR is age dependent and can be calculated by formula: IHR = 118 – (0.57 × age) [21].

In healthy subjects, resting normal IHR must be higher than resting HR before denervation. A low resting HR, in the presence of a normal IHR, proves an extrinsic origin of the sinus bradycardia. A very low IHR (below 60 bpm) is found in patients with intrinsic SND [22].

After the autonomic blockade, the rhythm is more stable, and the results of SN function tests are more reproducible. Some authors consider that vagal denervation with atropine alone can provide the same information as whole autonomic denervation [15].

6.8.2 Effect on the Atrioventricular Node

6.8.2.1 Atrioventricular Node Automaticity

Atropine stimulates the junctional pacemaker in normal subjects as well as patients with SN disease. The increase in heart rate is seen immediately after the administration of the drug and in some cases can be sustained, even for several hours [1].

6.8.2.2 Atrioventricular Node Conduction

In patients with normal AV conduction, atropine has a well-known positive dromotropic effect [1].

In patients with a grade 1 AV block, atropine ameliorates the conduction, but the effect may vary depending on the pathology and the level of block.

In patients with grade 2 type I AV blocks, atropine usually ameliorates the conduction. However, in patients with grade 2 type II AV blocks, atropine has the potential to aggravate the degree of (infranodal) block by facilitation of conduction at AV node level.

6.8.3 Effect on the His-Purkinje and Ventricular Muscle

Atropine has no direct effect on infranodal conduction but, by facilitating AV conduction, can play a role in assessing His-Purkinje effective refractory periods (ERP) in patients whose long AV node ERP in basal state precludes the assessment of ERP at inferior levels [23].

6.8.4 Atropine in Atrioventricular Nodal Reentry Tachycardia (AVNRT)

In most patients, typical AVNRT is induced after the conduction jump, when a critical prolongation of AH interval is produced in the slow anterograde pathway, to permit retrograde conduction in the fast pathway, and this can be reproducibly demonstrated by means of rapid atrial pacing or programmed stimulation with introduction of atrial premature stimuli.

In some patients, neither the usual nor more sophisticated protocols, such as stimulating at multiple cycle lengths, at multiple atrial sites and with multiple extrastimuli, are sufficient to induce the tachycardia. Assuming that electrophysiological properties of the nodal conduction are influenced by autonomic innervation, fluctuation of the adrenergic or parasympathetic tone presumably plays a role in creating conditions for AV node reentry. Consequently, drugs that facilitate conduction of one or other nodal pathways can be used to unmask the dual AV node physiology and/or to induce the arrhythmia. In this setting, the effect of isoprenaline and atropine (the most commonly used drugs), as well as propranolol, digitalis, and calcium blockers (less used), has been cited. Due to its anticholinergic effect, atropine shortens the refractory periods of the AV node and in some cases facilitates AVNRT induction by providing the necessary balance between conduction and refractoriness within the AV nodal reentrant pathways [24]. Therefore, atropine is used in the electrophysiology laboratory for unmasking the dual AV physiology, induction of the reentry, and facilitation of more sustained forms of tachycardia, (Figs. 6.6 and 6.7), but its effects are not constant.

Neuss et al. [25] studied 14 patients with paroxysmal supraventricular tachycardia (PSVT) and dual AV physiology. A total of 9 patients received 0.5 mg of atropine. After administration, the refractory periods of both fast and slow pathway were reduced; in some patients the effect of the reduction in ERP of the fast pathway was the disappearance of the dual AV node physiology.

Wu et al. [26] studied 14 patients with documented PSVT and demonstrated that atropine facilitates the induction of tachycardia by shortening the anterograde slow pathway conduction and retrograde fast pathway. In the same time, the arrhythmia was more sustained due to the shortening of conduction times in both pathways. Of note, in this study the conduction in the anterograde fast pathway was also facilitated by atropine, and this has demonstrated the potential to prevent antegrade slow pathway ERP with the atrial extrastimulus technique and, as a consequence, to mask dual AV node physiology.

The effect of autonomic blockade (AB) with atropine 0.04 mg/kg and propranolol 0.2 mg/kg was studied by Lin et al. [27] in 17 consecutive patients with typical AVNRT, with or without isoproterenol dependency for induction of arrhythmia. They demonstrated a tendency of either masking of the dual physiology (when present at baseline) or inducing tachycardia after AB. They postulated that this effect of AB was linked to a greater magnitude of shortening of the refractory periods of fast comparing to slow pathway and would be an argument for the dominance of parasympathetic control of dual AV physiology.

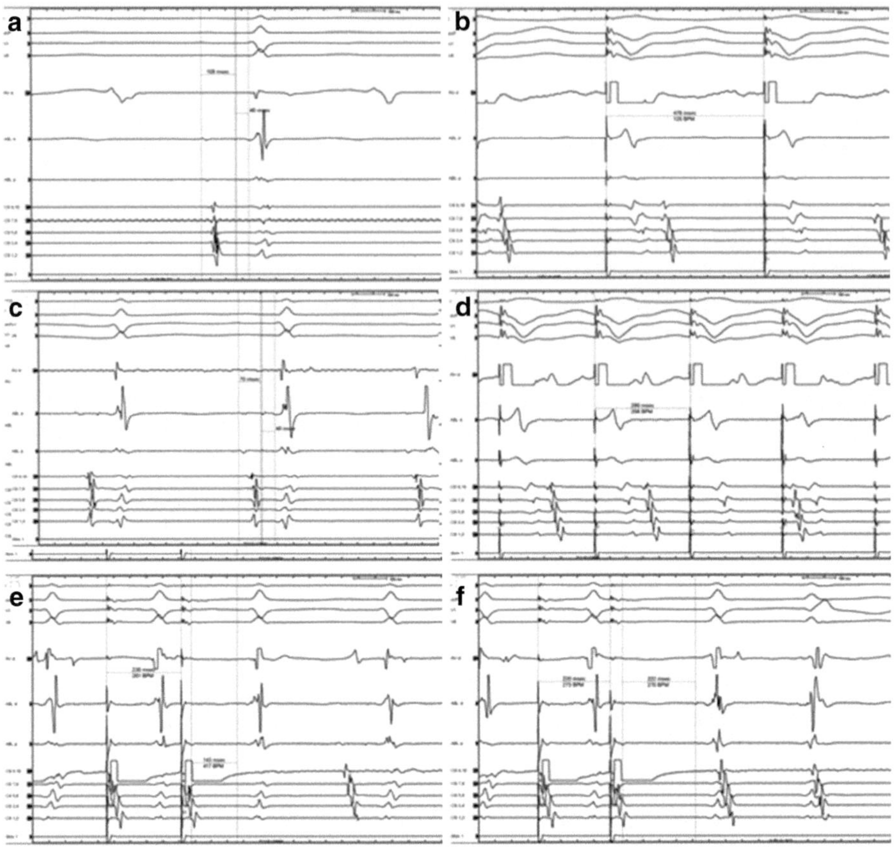

Fig. 6.6 Demonstration of value of atropine in the diagnosing of typical slow-fast AVNRT in a patient with poor retrograde conduction and no anterograde dual AVN node physiology in the basal state. The patient is a 30-year-old female patient with recurrent undocumented episodes of palpitations. Her resting 12-lead ECG was normal. During the electrophysiologic study in the basal state, the AH and HV intervals were normal, 106 ms and 40 ms, respectively (**a**). The anterograde Wenckebach point was 318 ms and the max AH interval 278 ms (not shown). Rapid atrial pacing or atrial extrastimulation failed to reveal dual AVN physiology. The retrograde conduction was concentric but rather "weak" with a retrograde Wenckebach point of 478 ms (**b**). In this patient, atropine administration ameliorated both the fast pathway anterograde conduction, with a decrease of the AH interval from 106 to 70 ms (**c**), and the retrograde conduction, with a decrease of the Wenckebach point from 318 to 290 ms (**d**) and also allowed demonstration of the dual AVN physiology with typical AV jump and induction of slow-fast form of AVNRT by atrial extrastimulus (**e**, **f**)

Stellbrink et al. [28] randomized 80 patients to receive either atropine (0.01 mg/kg) or isoproterenol (0.5–1 μg/kg/min) after baseline assessment of AV conduction. They have demonstrated that inducibility of AVNRT was reduced after atropine administration compared to isoproterenol and concluded that this effect is due to a more powerful effect of isoprenaline in facilitation of slow pathway conduction.

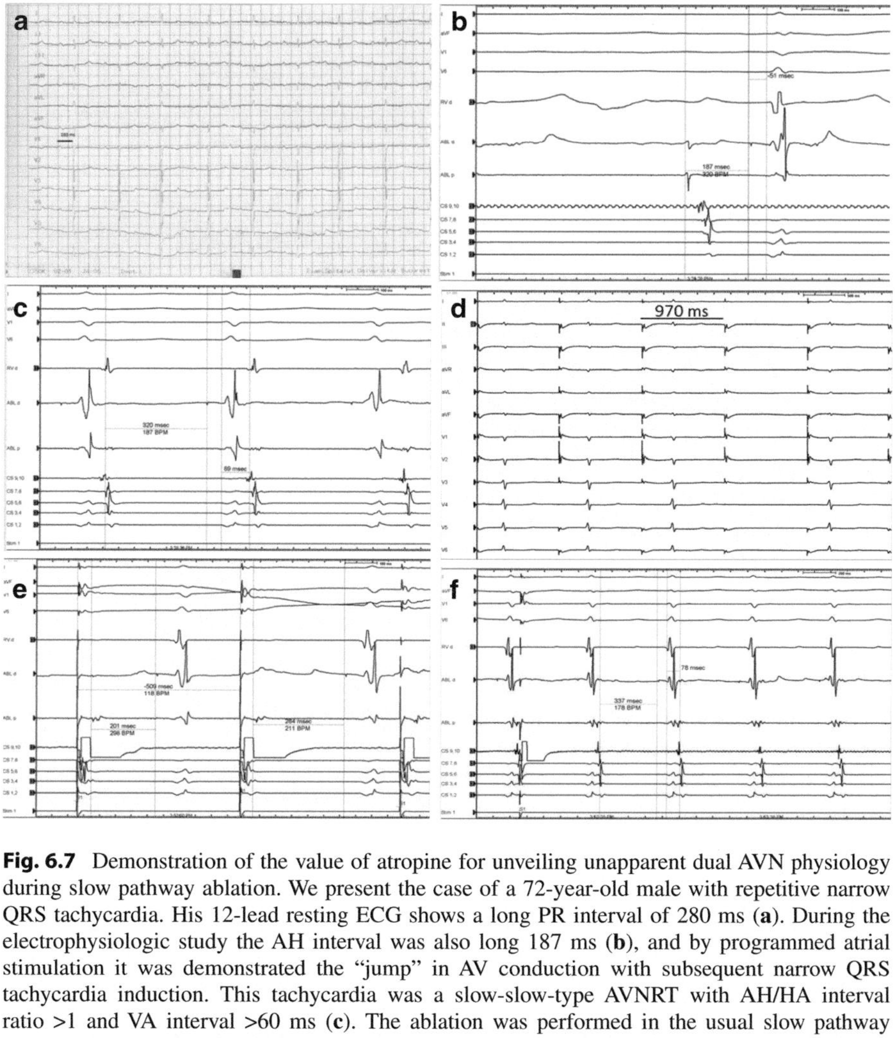

Fig. 6.7 Demonstration of the value of atropine for unveiling unapparent dual AVN physiology during slow pathway ablation. We present the case of a 72-year-old male with repetitive narrow QRS tachycardia. His 12-lead resting ECG shows a long PR interval of 280 ms (**a**). During the electrophysiologic study the AH interval was also long 187 ms (**b**), and by programmed atrial stimulation it was demonstrated the "jump" in AV conduction with subsequent narrow QRS tachycardia induction. This tachycardia was a slow-slow-type AVNRT with AH/HA interval ratio >1 and VA interval >60 ms (**c**). The ablation was performed in the usual slow pathway position. After one RF application there was an apparent complete disappearance of the conduction in the slow pathway, with anterograde Wenckebach point at 970 ms (**d**), maximum AH interval of 220 ms, and no AV "jump." Atropine administration (1 mg IV) unmasked the "dormant" dual AVN physiology facilitating sustained 1/1 conduction in the slow pathway (**e**) and induction of the AVNRT (**f**). Further application of RF abolished the sustained conduction in the anterograde slow pathway and rendered the tachycardia noninducible

6.8.5 Use of Atropine in Preexcitation Syndromes

Atropine can aid in the diagnosis of preexcitation by demonstrating the concept of fusion. The degree of preexcitation will be reduced, and the QRS will be consequently narrower after atropine administration owing to enhanced AV nodal conduction [15, 28, 29].

6.8.6 Use of Atropine in Ventricular Tachycardia (VT)

In selected cases, atropine (alone or in association with adrenaline or theophylline) can be used in an attempt to induce VT, when more common methods, as programmed ventricular stimulation and isoproterenol administration, fail [15].

6.8.7 Use of Atropine in Assessing Cardiac Autonomic Denervation Procedures

Ganglionated plexus (GP) ablation is an emerging technique for the treatment of patients with severe symptomatic functional bradycardia, as an alternative to cardiac pacing. Basically, the GP is targeted by different methods for ablation, and the acute response is appreciated by modification of various SN and AV conduction parameters, in the basal state and after atropine administration. Today, there still exists no consensus on the immediate end points of the procedure [30], but generally, the final parameters (with significance of vagal denervation) must not vary before and after administration of a standard dose of atropine (0.04 mg/kg).

References

1. Schweitzer P, Mark H. The effect of atropine on cardiac arrhythmias and conduction. Part 1. Am Heart J. 1980;100:255–61.
2. Robinson GC, Draper G. Rhythmic changes in the human heart. Heart. 1912;4:97–106.
3. Wilson FN. The production of atrioventricular rhythm in man after the administration of atropine. Arch Intern Med. 1915;16:989–98.
4. Thomas M, Woodgate D. The effect of atropine on bradycardia and hypotension in acute myocardial infarction. Br Heart J. 1966;28:409–13.
5. Harris A, Bluest R. Treatment of slow heart rate following acute myocardial infarction. Br Heart J. 1966;28:631–7.
6. Adgey AAJ, Geddes JA, Mulholland HC, Keegan DAJ, Pantridge JF. Incidence, significance and management of early bradyarrhythmias complicating acute myocardial infarction. Lancet. 1968;23:1097–101.
7. Shillingford J, Thomas M. Treatment of bradycardia and hypotension syndrome in patients with acute myocardial infarction. Am Heart J. 1968;75:843–4.
8. Harvey RD, Belevych AE. Muscarinic regulation of cardiac ion channels. Br J Pharmacol. 2003;139:1074–84.
9. Montano N, Cogliati C, Porta A, Pagani M, Malliani A, Narkiewicz K, et al. Central vagotonic effects of atropine modulate spectral oscillations of sympathetic nerve activity. Circulation. 1998;98:1394–9.
10. Hinderling PH, Gundert-Remy U, Schmidlin O, Heinzel G. Integrated pharmacokinetics and pharmacodynamics of atropine in healthy humans. II: pharmacodynamics. J Pharm Sci. 1985;74:711–7.
11. Adams RG, Verma P, Jackson AJ, Miller RL. Plasma pharmacokinetics of intravenously administered atropine in normal human subjects. J Clin Pharmacol. 1982;22:477–81.
12. https://www.drugs.com/pro/atropine.html
13. Van der Meer MJ, Hundt HK, Müller FO. The metabolism of atropine in man. J Pharm Pharmacol. 1986;38:781–4.

14. Kahn G. Cardiac drug therapy. 7th ed. Totowa: Humana Press; 2007. p. 198.
15. Josephson ME. Sinus node function. In: Josephson ME, editor. Clinical cardiac electrophysiology. Philadelphia: Wolters Kluwer/Lippincott Williams &Wilkins; 2008. p. 71.
16. Craig FN. Effects of atropine, work and heat on heart rate and sweat production in man. J Appl Physiol. 1952;4:826–33.
17. Mandel WJ, Hayakawa H, Allen HN, Danzig R, Kermaier AI. Assessment of sinus node function in patients with the sick sinus syndrome. Circulation. 1972;46:761–9.
18. Mandel WJ, Laks MM, Obayashi K. Sinus node function; evaluation in patients with and without sinus node disease. Arch Intern Med. 1975;135:388–94.
19. Morton HJV, Thomas ET. Effect of atropine on heart rate. Lancet. 1958;2:1313.
20. Reiffel JA, Bigger JT, Giardina EG. "Paradoxical" prolongation of sinus nodal recovery time after atropine in the sick sinus syndrome. Am J Cardiol. 1975;36:98–104.
21. Dhingra RC, Amat-Y-Leon F, Wyndham C, Denes P, Wu D, Miller RH, et al. Electrophysiologic effects of atropine on sinus node and atrium in patients with sinus nodal dysfunction. Am J Cardiol. 1976;38:848–55.
22. Opthof T. The normal range and determinants of the intrinsic heart rate in man. Cardiovasc Res. 2000;1(45):173–6.
23. Sethi KK, Balachandar J, Jaishankar S, Gupta MP. Differential effects of autonomic blockade on sinus and supraventricular tachycardia and dual atrioventricular atrioventricular nodal function in normals and in intrinsic sinus node dysfunction. Int J Cardiol. 1986;12:233–42.
24. Bissett JK, de Soyza ND, Kane JJ, Murphy ML. Electrophysiology of atropine. Cardiovasc Res. 1975;9:73–80.
25. Neuss H, Schlepper M, Spies HF. Effects of heart rate and atropine on 'dual AV conduction'. Br Heart J. 1975;37:1216–27.
26. Wu D, Denes P, Bauernfeind R, Dhingra RC, Wyndham C, Rosen KM. Effects of atropine on induction and maintenance of atrioventricular nodal reentrant tachycardia. Circulation. 1979;59:779–88.
27. Lin JL, Stephen Huang SK, Lai LP, Ko WC, Tseng YZ, Lien WP. Clinical and electrophysiologic characteristics and long-term efficacy of slow-pathway catheter ablation in patients with spontaneous node pathways without inducible tachycardia. J Am Coll Cardiol. 1998;31:855–60.
28. Stellbrink C, Diem B, Schauerte P, Brehmer K, Schuett H, Hanrath P. Differential effects of atropine and isoproterenol on inducibility of atrioventricular nodal reentrant tachycardia. J Interv Card Electrophysiol. 2001;5:463–9.
29. Ito M. The effects of atropine on atrio-ventricular conduction in patients with Wolff-Parkinson-White syndrome. Studies with His bundle electrogram. Jpn Circ J. 1976;40:1285–300.
30. Rivarola EW, Hachul D, Wu T, Pisani C, Hardy C, Raimundi F, et al. Targets and end points in cardiac autonomic denervation procedures. Circ Arrhythm Electrophysiol. 2017;10:1–9.

How to Induce Arrhythmias with Noradrenaline

7

Ispas Alexandru

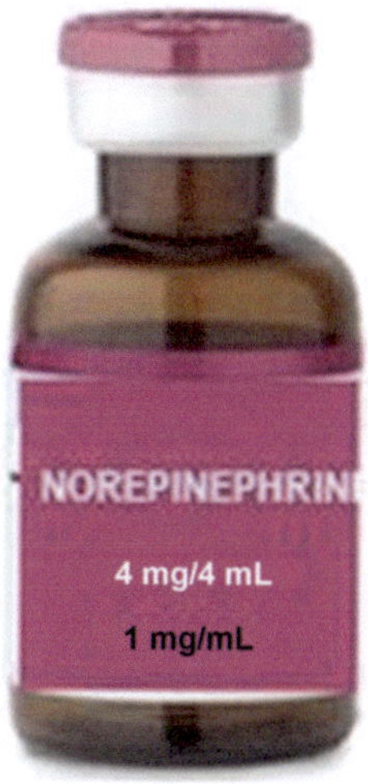

7.1 Introduction

Noradrenaline is a precursor of adrenaline, secreted by the adrenal part of the medulla, and like dopamine is a neurotransmitter found in the brain and other parts of the body. Noradrenaline is important in the response and recovery after injury or serious illness. Furthermore, significant plasma levels are seen in patients with different types of shock, severe hypoxemia, or cardiac failure.

I. Alexandru (✉)
Cardiology Department, Centre Hospitalier Annecy-Genevois, Annecy, France

G. Cismaru (ed.), *Arrhythmia Induction in the EP Lab*,
https://doi.org/10.1007/978-3-319-92729-9_7

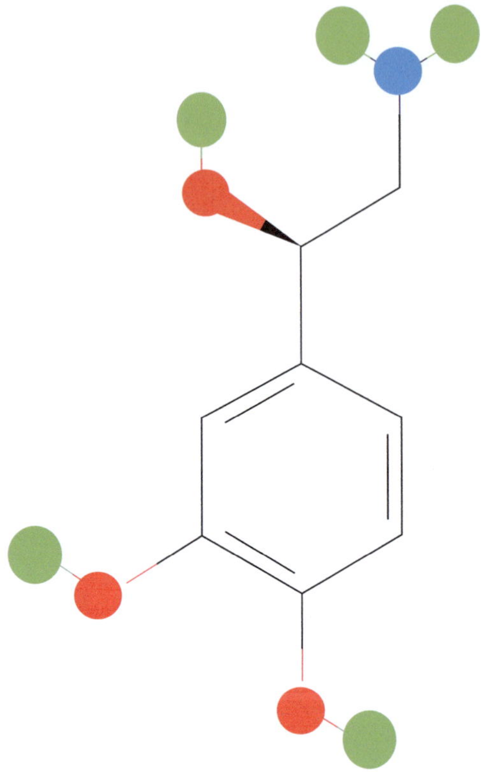

Fig. 7.1 Chemical structure of noradrenaline: $C_8H_{11}NO_3$

Cellular response to noradrenaline is mediated by receptors that are located on target cells. It has the potential of facilitating cardiac arrhythmias by β1 receptor stimulation, alpha-1 receptor stimulation, or increase in myocardial oxygen consumption.

The name noradrenaline is an acronym of "nitrogen ohne radikal" which indicates the absence of a methyl group from the chemical formula (Fig. 7.1). After noradrenaline was separated from adrenaline and demonstrated as the principal sympathomimetic neurotransmitter, it started to be clinically used [1].

7.2 Pharmacology

When noradrenaline is injected intravenously, a rise in the blood pressure occurs rapidly. The effect is of short duration. After 1–2 min of discontinuation of infusion, the pressor response disappears. The metabolism of noradrenaline is based on COMT (cathecol-O-methyltransferase) and MAO (monoamine oxidase). The major metabolites are called metanephrines and vanillylmandelic acid.

7.3 Mechanism of Action

Noradrenaline is a sympathicomimetic with a double effect on the α and β postsynaptic receptors. Furthermore, sympathetic stimulation by noradrenaline leads to adrenaline secretion from the medulosuprarenal gland, with further effect on the α and β receptors. On the other hand, the effects of noradrenaline are different of those of adrenaline, even though they both act on the α and β receptors.

Adrenergic receptors are responsible for cardiac output, peripheral vasculature's tone, and control of the blood pressure. Alpha-1 receptors are located in vascular smooth cells, but they also can be found in the heart, kidneys, uterus, and liver. β1 are found in cardiac myocytes and β2 in vascular and respiratory smooth cells [2].

Stimulation of α1 receptors with **noradrenaline** results in peripheral vasoconstriction. Stimulation of α1 cardiac receptors increases myocardial contractility without any increase in heart rate. β1 receptor stimulation leads to increase in heart rate and myocardial contractility. Stimulation of β2 also results in positive inotropic and chronotropic effects. By stimulating β2 receptors, norepinephrine results in vasodilation and bronchodilation (Tables 7.1 and 7.2).

Noradrenaline's electrophysiologic effects are similar to those of adrenaline. It increases the sinus node automaticity; it decreases the action potential duration of the sinus node but with no effect on the amplitude of the action potential or the resting membrane potential [3].

At the atrial level noradrenaline increases the resting membrane potential and also decreases the conduction time through the AV node [4]. At ventricular level, it increases automaticity, amplitude, and duration of the action potential as well as the resting membrane potential. It also decreases the absolute refractory period of both right and left ventricle [5]. Like adrenaline, it decreases the ventricular fibrillation threshold.

Table 7.1 Adrenergic agonist activity of Noradrenaline compared to other vasopressor agents

Agent	Alpha 1	Alpha 2	Beta 1	Beta 2
Dopamine	++	+/−	++++	++
Adrenaline	++++	+/−	++++	+++
Noradrenaline	+++	+/−	++++	+
Phenylephrine	++	0	0	0

Table 7.2 Electrophysiologic effects of Norepinephrine compared to other vasopressor agents

Agent	APD	CL	AERP	AH	HV	AVERP	VERP
Dopamine	↓	↓		↓			↔
Epinephrine	↑	↓	↓	↓	↓	↓	↓
Noradrenaline	↓	↓	↓	↓	↓	↓	↓
Phenylephrine	↔	↓					↑

APD action potential duration; *CL* sinus cycle length; *AERP* atrial effective refractory period; *AH* AH interval; *HV* HV interval; *AVERP* atrioventricular effective refractory period; *VERP* ventricular refractory period

7.4 Doses of Noradrenaline

One vial of 4 mL contains 8 mg noradrenaline tartrate which is equivalent to 4 mg noradrenaline base.

The initial dose is 0.05 µ/kg/min which is further increased to a maximum dose of 1 µg/kg/min (Fig. 7.2, Tables 7.3, 7.4 and 7.5).

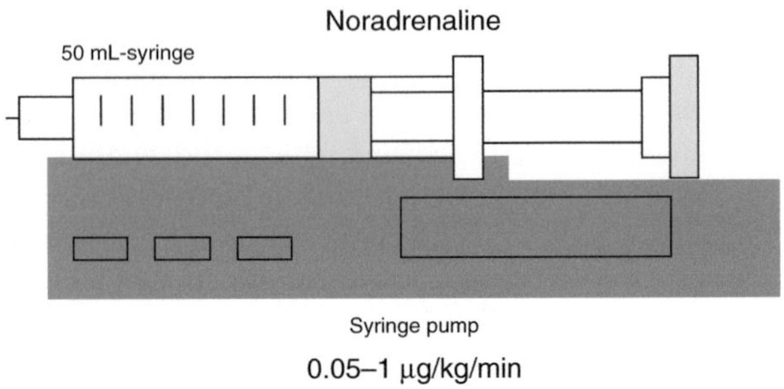

Fig. 7.2 Noradrenaline infusion dose

Table 7.3 Noradrenaline dose on time unit

Time	Dose
Start	0.05 µg/kg/min
3 min	0.1 µg/kg/min
6 min	0.25 µg/kg/min
9 min	0.5 µg/kg/min
12 min	1 µg/kg/min

Table 7.4 Protocol for administration of Noradrenaline

Infusion preparation	
1.	The patient should not eat for at least 8 h before the electrophysiological study
2.	Patients are instructed to stop medication five times the half time prior to electrophysiological testing unless instructed otherwise
3.	An IV is started using aseptic technique
4	An infusion pump is necessary for noradrenaline administration
5.	Noradrenaline can be diluted with either glucose 5% or sodium chloride 0.9%
6.	Add 2 mL of noradrenaline into 50 cc syringe with 48 mL of NaCl 0.9% or glucose 5% into the syringe
7.	The final concentration of noradrenaline is 40 mg/L noradrenaline base (equivalent to 30 mg/L noradrenaline tartrate)
8.	Noradrenaline is started at a dose of COS µg/kg/min
9.	Continuous monitoring of EKG, heart rate, and 1-min assessment of blood pressures and symptoms for each infusion stage
10.	Noradrenaline infusion is increased by 0.05 µg/kg/min at 3-min intervals up to a maximum of 1 µg/kg/min
11.	Atropine 1–3 mg may be used if clinical arrhythmia cannot be induced by programmed atrial or ventricular stimulation

Table 7.5 How to prepare the Noradrenaline infusion

Understanding the numbers:

You are ordered to give 0.1 µg/kg/min of Noradrenaline to your 58-year-old, 70 kg patient to facilitate AVNRT induction

1. The Noradrenaline vial has 4 mg/4 mL
2. You should take 1 vial of Noradrenaline in the 50 mL syringe with 46 mL of serum (0.9% NaCl)
3. This gives a concentration of 4 mg/50 mL meaning 0.08 mg/mL or 80 µg/mL
4. 0.1 µg x 70 kg/min=7 µg/min should be injected, equivalent of 420 µg/h
5. We will start the perfusion rate: 420:80=5.25 mL/h

7.5 Preparation

Infusion preparation

1. The patient should not eat for at least 8 h before the electrophysiological study.
2. Patients are instructed to stop medication 5 times the half time prior to electrophysiological testing unless instructed otherwise.
3. An IV is started using aseptic technique.
4. An infusion pump is necessary for noradrenaline administration.
5. Noradrenaline can be diluted with either glucose 5% or sodium chloride 0.9%.
6. Add 2 mL of noradrenaline into 50 cc syringe with 48 mL of NaCl 0.9% or glucose 5% into the syringe.
7. The final concentration of noradrenaline is 40 mg/L noradrenaline base (equivalent to 89 mg/L noradrenaline tartrate).
8. Noradrenaline is started at a dose of 0.05 mcg/kg/min.
9. Continuous monitoring of EKG, heart rate, and 1-min assessment of blood pressures and symptoms for each infusion stage.
10. Noradrenaline infusion is increased by 0.05 µg/kg/min at 3-min intervals up to a maximum of 1 µg/kg/min.
11. Atropine 1–3 mg may be used if clinical arrhythmia cannot be induced by programmed atrial or ventricular stimulation.

7.6 Side Effects

Intravenous inotropic agents promote increased myocardial contractility via elevation of myocyte calcium concentrations, a mechanism that is also known to promote the development of cardiac arrhythmias (Tables 7.6, 7.7 and 7.8).

Table 7.6 Side effects of Noradrenaline

Side effects of noradrenaline
• Necrosis in case of extravasation
• Sweating
• Vomiting
• Photophobia
• Anxiety
• Dyspnea
• Gangrene

Table 7.7 Contraindications for Noradrenaline

Contraindications for noradrenaline
• Hypersensitivity to noradrenaline or
• Sodium chloride
• Sodium hydroxide
• Hydrochloric acid
• Caution in patients who exhibit profound hypoxia or hypercarbia
• Caution in patients with coronary, mesenteric, or peripheral vascular thrombosis as it may increase the ischemia and extend the area of infarction

Table 7.8 Pro-arrhythmic potential of Noradrenaline compared to other vasopressors

Agent	SB	ST	SVT	VT
Dopamine	0	+++	++	+++
Adrenaline	0	++	++	+++
Noradrenaline	0	++	++	+++
Phenylephrine	+	0	0	0s

SB sinus bradycardia; *ST* sinus tachycardia; *SVT* supraventricular tachycardia; *VT* ventricular tachycardia

7.7 Electrophysiologic Effects

7.7.1 Animal Studies

The effects of noradrenaline are very similar to those of adrenaline. In sinus node preparations from the guinea pig, high doses of noradrenaline increase the automaticity of the sinus node, decrease the duration of the action potential, and have no effect on the action potential amplitude or resting membrane potential [6]. In atrial myocytes from rabbits, norepinephrine increases the resting membrane potential and increases the speed through the AV node [7].

In canine ventricular myocytes noradrenaline increases the automaticity, the action potential duration and amplitude, and the resting membrane potential. On the other hand, noradrenaline decreases the ventricular refractory period [8].

Using isolated innervated rabbit hearts, Winter et al. showed that in presence of parasympathetic activation (vagus nerve stimulation), rapid infusion of noradrenaline may trigger torsade de pointes, suggesting that concomitant autonomic activation (or "autonomic conflict") is required for the adrenergic facilitation of long QT syndrome (LQTS)-associated arrhythmias*.

Noradrenaline can be used for ventricular tachycardia induction in animals as described by Friedrichs [9].

Wax [10] and Papp [11] showed that noradrenaline increases the ventricular fibrillation threshold in dogs.

In rats, Sharma showed that noradrenaline in doses of 0.04–1.0 μg/kg/min reduces the duration of induced ventricular fibrillation to the time of sinus rhythm [12].

In summary, the abovementioned studies on animal preparations show that noradrenaline increases the automaticity of the sinus node and atrial and ventricular myocardium. It also enhances conduction through the AV node and decreases the ventricular refractory period.

7.7.2 Human Studies

The effects of noradrenaline infusion during electrophysiological study were evaluated by Weiss et al. who showed on 21 patients that a dose of 25 ng/kg/min led to a blood concentration of 298–708 pg/mL. The effect was increase in the blood pressure, increase in the cycle length from 831 to 908 ms, increase of the sinus node recovery time from 296 to 447 ms, increase in the AH interval from 87 to 93 ms, increase in the Wenckebach point from 380 to 420 ms, increase of the atrial refractory period from 225 to 235 ms, and increase of the ventricular refractory period from 221 to 228 ms. There were no modifications of the HV interval nor AV node refractory period after noradrenaline infusion [2]. The blood concentration of adrenaline was unchanged after noradrenaline infusion. We have to mention that the dose used in the electrophysiology laboratory was higher than the usual dose for clinical use which is 2–4 μg/min. The injected dose of noradrenaline led to plasmatic concentrations considered "physiological," similar to those obtained during orthostatism, public speaking, mild hypoglycemia, smoking, myocardial infarction, dental extraction, surgery, and submaximal stress test. Weiss et al. explained these values by sinus node and atrioventricular node inhibition by two mechanisms: first, a direct stimulation of the α receptors, even though adrenaline is administrated intravenously, there is no inhibition of the sinus node and atrioventricular node. The second mechanism is an indirect rise of the vagal tone, by blood pressure increase and baroreceptors activation, which was further confirmed by Goldenberg et al. who showed that atropine reverses the inhibition effect on the sinus node determined by noradrenaline [13].

On the other hand, three prospective unblinded and uncontrolled studies evaluated noradrenaline in doses of 0.01–1.5 μg/kg/min for periods of time ranging from 1 to 240 h. No patients experienced severe arrhythmias [14].

In a multicenter randomized trial comparing dopamine with noradrenaline in patients with shock, De Baker et al. [15] found less arrhythmic events in patients receiving noradrenaline infusion in doses of 0.02–0.19 μg/kg/min: atrial fibrillation in 90 patients (11%), ventricular tachycardia in 8 patients (1%), and ventricular fibrillation in 4 patients (0.5%).

References

1. von Euler US, Harnberg U. L-noradrenaline in the suprarenal medulla. Nature. 1949;163:642–63.
2. Weiss R, Knight BP, Bahu M, Zivin A, Souza J, Goyal R, Daoud E, Man C, Strickberger A, Halter JB, Morady F. Cardiac electrophysiologic effects of norepinephrine in human beings. Am Heart J. 1998;135:945–51.
3. Brown RA, Carpentier RG. Alpha-adrenoreceptor- mediated effects of norepinephrine on the guinea pig sinus node. J Electrocardiol. 1988;21:213–7.
4. Gilbert JL, Lange G, Polevoy I, Brooks CM. Effects of vasoconstrictor agents on cardiac irritability. J Pharmacol Exp Ther. 1958;123:9–15.
5. Hauswirth O, Noble D, Tsien RW. Adrenaline: mechanism of action on the pacemaker potential in cardiac Purkinje fibers. Science. 1968;162:916–7.
6. Brown RA, Carpentier RG. Alpha-adrenoceptor-mediated effects of norepinephrine on the guinea pig sinus node. J Electrocardiol. 1988;21:213–7.
7. Fedida D, Shimoni Y, Giles WR. A novel effect of norepinephrine on cardiac cells is mediated by alpha adrenoceptors. Am J Physiol. 1989;256:H151XI-4.
8. Lukas A, Ferrier GR. Arrhythmic effects of norepinephrine in a model of cardiac ischemia and reperfusion. Can J Physiol Pharmacol. 1989;67:765–71.
9. Friedrichs GS, Merrill GF. Adenosine desaminase and adenosine attenuate ventricular arrhythmias caused by norepinephrine. Am J Phys. 1991;260:H979–84.
10. Wax SD, Webb WR, Ecker RR, Sugg WL. Evaluation of antifibrillatory agents and catecholamines by a physiologic method. Surg Forum. 1968;19:J49–51.
11. Papp JG, Szekeres L. Analysis of the mechanism of adrenergic actions on ventricular vulnerability. Eur J Pharmacol. 1968;3:15–26.
12. Sharma AD, Waxman MB, Huerta F. Effect of adrenergic stimulation and blockade on ventricular defibrillation in the rat. J Pharmacol Exp Ther. 1983;227:716–22.
13. Goldenberg M, Pines KL, Baldwin EF, Green DG, Roh CE. The hemodynamic response of man to norepinephrine and epinephrine and relation to the problem of hypertension. Am J Med. 1948;5:792–806.
14. Desjars P, Pinaud M, Potel G, Tasseau F, Touze MD. A reappraisal of norepinephrine therapy in human septic shock. Crit Care Med. 1987;15:134–7.
15. De Baker D, Biston P, Devriendt J, et al. Comparison of dopamine and norepinephrine in the treatment of shock. N Engl J Med. 2010;362:779–89.

How to Induce Arrhythmias with Dobutamine

8

Abu Walid Nasra

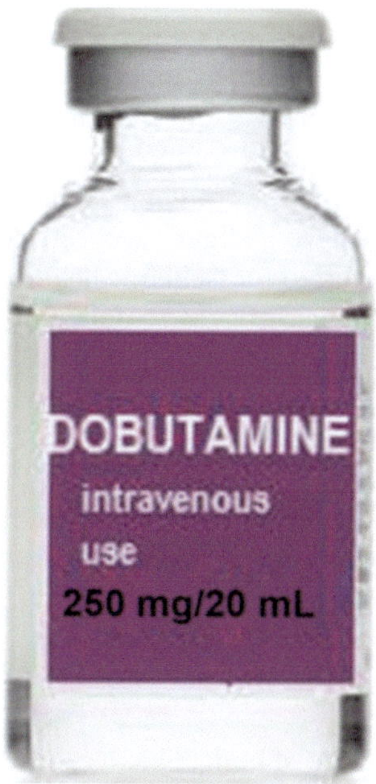

8.1 Introduction

In 1975, Tuttle and Mills published an article on a new catecholamine discovered in 1970: dobutamine. Tuttle, a pharmacologist, and Mills, a chemist, modified the structure of isoprenaline to obtain a new drug. Isoprenaline through its action on β

A. W. Nasra (✉)
Mazor Health Center, Acre, Northern District, Israel

G. Cismaru (ed.), *Arrhythmia Induction in the EP Lab*,
https://doi.org/10.1007/978-3-319-92729-9_8

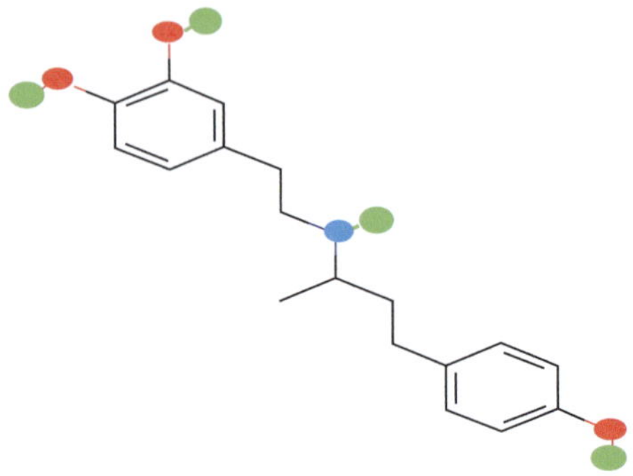

Fig. 8.1 Chemical structure of Dobutamine: $C_{18}H_{23}NO_3$

adrenergic receptors increases the cardiac contractility with the cost of an increased heart rate and proarrhythmic effect associated with an increase in blood pressure. They wanted to obtain a new compound with a reduced risk of arrhythmia and less effect on heart rate. After removing the hydroxyl group from isoproterenol, the new compound had a positive chronotropic and inotropic effect. By substitution of the side chains (Fig. 8.1), they obtained a compound with less chronotropic effect that reduced the arrhythmogenic properties of the product [1, 2].

Dobutamine was found to have a slight vasopressor activity and only at higher doses, a compound that is preferred for ischemic heart disease where blood pressure should be lower [3]. In less than 10 years, dobutamine was synthesized, tested through preclinical tests and also approved for clinical use. Dobutamine is a chemical precursor of isoproterenol and can be an alternative for inducing arrhythmia during programmed atrial and ventricular stimulation. Dobutamine was designed as an inotropic agent for patients with congestive heart failure. Atrial and ventricular arrhythmias are only side effects of the drug, but this side effects can become advantages for the electrophysiologists when isoprenaline is unavailable in their lab.

8.2 Pharmacology

Dobutamine is a racemic mixture of both (+) and (−) isomers. It is given as an intravenous infusion because it disappears very rapidly from the systemic circulation. Dobutamine has a half-life of 2 min.

Dobutamine is catalyzed by COMT, and the major products of excretion are 3-O-methyl-dopamine and conjugates of dobutamine, especially with sulfate [4].

8.3 Mechanism of Action

Dobutamine increases the automaticity of the sinoatrial node and decreases the refractory period of the atrium and atrio-ventricular node as well as the atrio-ventricular conduction time. It also decreases the refractory period of both ventricles, healthy or ischemic. It was also demonstrated in the study of Masoni et al. [5] that dobutamine decreases the sinus cycle length, the SNRT, and corrected SNRT. At the atrial and nodal levels, it decreases the AH interval, Wenckebach cycle length, AV nodal functional and effective refractory periods. Dobutamine has no effect on the HV interval. Besides an increase in the heart rate, dobutamine induces ventricular premature contractions in 3–15% of patients. Ventricular tachycardia can be induced by dobutamine infusion in a small percentage of patients. Ischemic patients, those with heart failure and known arrhythmias, are at higher risk of proarrhythmia [6].

Dobutamine has a direct ß1 and ß2 stimulation effect with a dose-related increase in heart rate, myocardial contractility, and blood pressure. Dobutamine increases regional myocardial blood flow but at a dose of 20 μg/kg/min induces coronary flow heterogeneity less than that induced by Dipyridamole or Adenosine. Since it has no action on dopamine receptors, is less prone to induce hypertension and induce norepinephrine release compared to dopamine. Dobutamine has also a mild ß2 agonist activity, making it useful as a vasodilator.

Dobutamine promotes an increase in myocardial contractility via increasing myocite calcium concentrations, a mechanism that is known to facilitate cardiac arrhythmias and can be used in the electrophysiological study to induce arrhythmias during programmed stimulation.

8.4 Doses of Dobutamine

Dobutamine is started at a dose of 5 μg/kg/min. The infusion is increased by 10 μg/kg/min at 3-min intervals up to a maximum of 50 μg/kg/min (Fig. 8.2, Tables 8.1 and 8.2).

The maximum dose that was used during dobutamine stress echocardiography test in the study of Mertes et al. was 50 μg/kg/min.

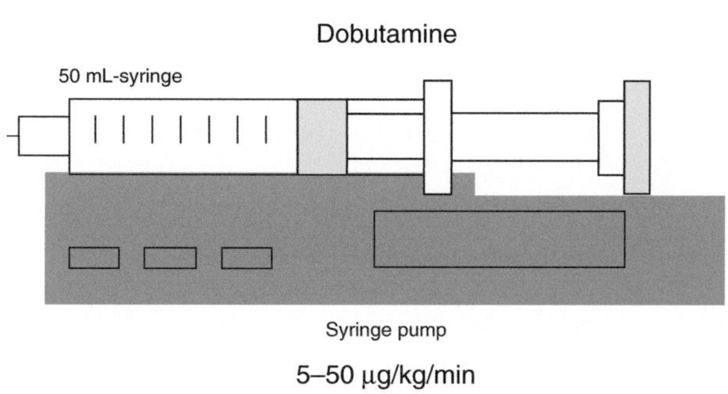

Fig. 8.2 Doses of Dobutamine

Table 8.1 Dobutamine dose on time unit

Time	Dose
Start	5 μg/kg/min
3 min	15 μg/kg/min
6 min	25 μg/kg/min
9 min	35 μg/kg/min
12 min	40 μg/kg/min
15 min	45 μg/kg/min
18 min	50 μg/kg/min

Table 8.2 Protocol for administration of Dobutamine

1.	The patient should not eat for at least 8 h before the electrophysiological study
2.	Patients are instructed to stop medication five times the half time prior to electrophysiological testing unless instructed otherwise
3.	An IV is started using aseptic technique
4.	An infusion pump is necessary for dobutamine administration
5.	Add 8 mL of 250/20 mL dobutamine into 50 cc syringe with 42 mL of NaCl 0.9% into the syringe
6.	Dobutamine is started at a dose of 5 μg/kg/min
7.	Continuous monitoring of EKG, heart rate, and 1-min assessment of blood pressures and symptoms for each infusion stage
8.	Dobutamine infusion is increased by 10 μg/kg/min at 3-min intervals up to a maximum of 40 μg/kg/min
9.	Atropine 1–3 mg may be used if clinical arrhythmia cannot be induced by programmed atrial or ventricular stimulation

8.5 Contraindications to Dobutamine During Electrophysiological Study

- Recent (<1 week) myocardial infarction
- Unstable angina
- Hemodynamically significant left ventricular (LV) outflow tract obstruction
- Severe aortic stenosis
- Uncontrolled hypertension (blood pressure
- >200/110 mmHg)
- Patients with aortic dissection or large aortic aneurysm
- Unwillingness to give consent (Table 8.3)

8.6 Dobutamine's Side Effects

Intravenous inotropic agents promote increased myocardial contractility via elevation of myocite calcium concentrations, a mechanism that is also known to promote the development of cardiac arrhythmias.

Table 8.3 Contraindications for Dobutamine

Contraindications for dobutamine during electrophysiological study
• Recent (<1 week) myocardial infarction
• Unstable angina
• Hemodynamically significant left ventricular (LV) outflow tract obstruction
• Severe aortic stenosis
• Uncontrolled hypertension (blood pressure >200/110 mmHg)
• Patients with aortic dissection or large aortic aneurysm
• Unwillingness to give consent

Table 8.4 Side effects of Dobutamine

Side effects of dobutamine
• High blood pressure
• Angina pectoris
• Myocardial infarction
• Coronary vasospasm
• Takotsubo cardiomyopathy
• Bronchospasm
• Nausea
• Headache

- Increase in systolic blood pressure
- Ventricular and supraventricular premature beats
- Angina, myocardial ischemia, coronary vasospasm
- Stress cardiomyopathy
- Exanthema, skin rash; phlebitis
- Anxiety
- Bronchospasm
- Nausea
- Headache
- Signs of poor perfusion: Cyanosis and pallor (Table 8.4)

8.7 Infusion Preparation

8.8 Electrophysiological Effects

8.8.1 Animal Studies

Nachar et al. studied the effects of dobutamine on 20 newborn piglets in dose of 2.5 and 7.5 μg/kg/min during 6 h and found improvement of the ventricular systolic and diastolic function within minutes after beginning of the infusion. The effect persisted for 6 h during the infusion time. Tachycardia was demonstrated for all 3 doses of dobutamine but affected the myocardial oxygen consumption only at the highest dose of 7.5 μg/kg/min [7].

Mc Goven et al. compared the effects of dobutamine 10 μg/kg/min and dopamine 10 μg/kg/min and adrenaline 0.1 μg/kg/min over 20 min on 10 young pigs with right ventricular injury. The injury was obtained by cryoablation of the RVOT and right ventricular free wall. After cryoablation in all animals, atrial pacing was performed for overdrive of junctional rhythms and slow rhythms. Although all three inotropes improved the function of the right ventricle both systolic and diastolic, only adrenaline increased pulmonary vascular resistance [8].

Ferrara et al. injected dobutamine 5, 10, and 15 μg/kg/min in 18 term and 16 premature piglets and compared with 5, 10, and 15 μg/kg/min dopamine. Heart rate increased moderately in a dose-dependent fashion in preterm animals as well as in term animals [9].

Dobutamine can increase the heart rate in a dose-related way: at doses of 5 μg/kg/min, it increases the heart rate in dogs with 17% and at doses of 50 μg/kg/min with 165% [10].

Bednarski et al. showed that dobutamine induces atrial premature beats in dogs at doses of 13.5 μg/kg/min or more. Premature ventricular contractions appear at doses of 17 μg/kg/min or more [11].

Parrat et al. infused dobutamine in 7 dogs with myocardial ischemia in doses of 10–24 μg/kg/min and documented 25–109 PVCs [12].

Kirlin et al. administrated dobutamine in 6 dogs with ligation and occlusion of the left circumflex coronary artery. One of the 6 dogs that received dobutamine presented ventricular tachycardia, but none of the 6 animals after placebo injection [13].

Ozaki et al. studied the effects of dobutamine during programmed ventricular stimulation. They performed EP study and induced NSVT in 2 of 10 dogs and sustained VT in 2 of 10 dogs with subacute myocardial infarction induced by ligation of the left descendent coronary artery. After 5 μg/kg/min of dobutamine, NSVT was inducible in 1 of the previously noninducible animals and 10 μg/kg/min permitted induction of sustained VT in this dog. The dose of 10 μg/kg/min also produced induction of NSVT in 1 dog and sustained VT in another dog that was uninducible in the basal state. Furthermore, 1 patient that had NSVT in the basal state developed sustained VT after dobutamine 10 μg/kg/min [14].

8.8.2 Human Studies

Vanegas et al. compared 144 patients with isoproterenol infusion with 140 patients with dobutamine infusion. They found no difference between the two groups in terms of arrhythmia induction. Their study lacks important information like the type of clinical arrhythmia, the percentage of induction, the statistical test used to compare both groups and the obtained p value. Despite this, it shows that when isoproterenol is not available in one EP lab, dobutamine can be used as an alternative for arrhythmia induction [15].

Leier et al. injected dobutamine in 25 patients with heart failure NYHA IV class. The dose was 2.5 μg/kg/min which was further escalated to 15 μg/kg/min and maintained for 72 h. The heart rate increased significantly with the 15 μg/kg/min dose [16] (Table 8.5).

One year later, the same author compared dobutamine with dopamine on 13 patients with heart failure NYHA III and IV. Dopamine significantly increased the heart rate at doses of 8 μg/kg/min or above, compared to dobutamine which increased contractility without any increase in the heart rate for doses below 10 μg/kg/min. Furthermore, dopamine when injected in doses higher than 4 μg/kg/min also increased the frequency of ventricular premature beats [17].

David et al. reported two cases with ventricular premature beats that received dobutamine 5 μg/kg/min. Holter monitoring showed a significant increase in the frequency and complexity of ventricular premature contractions, including multifocal PVCs and ventricular tachycardia. This observation further imposed reduction of dobutamine infusion to 2 or 4 μg/kg/min [18].

Gillespie et al. studied dobutamine's arrhythmogenic effects at doses of 15 μg/kg/min and higher in 16 patients with acute myocardial infarction. The number of premature ventricular contractions was higher than in the control group after dobutamine infusion: 130 PVCs for the control group after 13 h of dobutamine and 150 PVCs for the treated group after 13 h of dobutamine [19].

In the study of Berthe et al., Dobutamine was infused in 37 patients with myocardial infarction in doses of 5–40 μg/kg/min at 3 min interval. Ten patients received 40 μg/kg/min and they developed isolated PVCs [20]. In the case report of Guimon, a patient that was candidate for heart transplantation presented complex supraventricular arrhythmias and ventricular tachycardias associated with hemodynamic compromise at doses of 20–40 μg/kg/min of dobutamine [21].

Dies et al. demonstrated the influence of dobutamine on mortality in a study on 60 patients with refractory NYHA III and IV cardiac failure. The presence of 4 or more episodes of ventricular tachycardia under dobutamine was associated with mortality [22].

Arrhythmias can develop also during dobutamine stress echocardiography. Mertes et al. verified the safety of dobutamine on 1118 patients with dobutamine doses up to 50 μg/kg/min. Most of the patients, 65% had no arrhythmia during the test. Asymptomatic PVCs developed in 15% of patients, with a frequency of more than 6 beats/min. Atrial premature contractions were present in 8% of patients and asymptomatic NSVT occurred in 4% of patients. The longest episode of ventricular

Table 8.5 How to prepare the infusion of Dobutamine

Understanding the numbers:

You are ordered to give 20 μg/kg/min Dobutamine to your 58-year-old, 70 kg patient for induction of orthodromic reentrant tachycardia

1. The Dobutamine vial has 250 mg/20 mL
2. you should take1 vial of Dobutamine in the 50 mL syringe with 30 mL of serum (0.9% NaCl)
3. This gives a concentration of 250 mg in 50 mL meaning 5 mg/mL
4. 20 x 70 kg/min=1400 μg/min should be injected, equivalent of 84 mg/h
5. The infusion rate should be se at 17 mL/h

tachycardia was 20 s. One patient necessitated lidocaine to stop repeated episodes of NSVT. There were no deaths during the study and no recidives of VT during follow-up. In seven patients, second-degree AV block was present after dobutamine infusion and necessitated atropine administration. Also in 7 patients atrial fibrillation occurred after dobutamine infusion. Five converted spontaneously to sinus rhythm and 2 necessitated digoxin administration [23].

References

1. Tuttle RR, Millis J. Dobutamine: development of a new cathecolamine to selectively increase the cardiac contractility. Circ Res. 1975;36:185–96.
2. Tuttle RR. Inotropic and cronotropic effects of dobutamine and isoproterenol on cat atria and papillary muscles at different temperatures (abstr). Fed Proc. 1974;33:503.
3. Henzlova MJ, et al. ASNC imaging guidelines for nuclear cardiology procedures: stress protocols and tracers. J Nucl Cardiol. 2009.
4. Yan M, Webster LT Jr, Blumer JL. Kinetic interactions of dopamine and dobutamine with human catechol-*O*-methyltransferase and monoamine oxidase in vitro. J Pharmacol Exp Ther. 2002;301:1–7.
5. Masoni A, Alboni P, Malacarne C, et al. Effects of dobutamine on electrophysiological properties of the specialized conduction system in man. J Electrocardiol. 1979;12:361–70.
6. McNally EM. Can we do better than Dobutamine? Circ Res. 2013;113:355–7.
7. Nachar RA, Booth EA, Friedlich P, Borzage M, Soleymani S, et al. Dose-dependent hemodynamic and metabolic effects of vasoactive medications in normotensive, anesthetized neonatal piglets. Pediatr Res. 2011;70:473–9.
8. McGovern JJ, Cheifetz IM, Craig DM, Bengur AR, Quick G, et al. Right ventricular injury in young swine: effects of catecholamines on right ventricular function and pulmonary vascular mechanics. Pediatr Res. 2000;48:763–9.
9. Ferrara JJ, Dyess L, Peeples GL, Christenberry P, Roberts WS, et al. Effects of dopamine and dobutamine on regional blood flow distribution in the neonatal piglet. Ann Surg. 1995;221:531–42.
10. Leighton KM, Bruce C. Dobutamine and general anaesthesia: a study of the response of arterial pressure, heart rate and renal blood flow. Can Anaesth Soc J. 1976;23:176–84.
11. Bednarski RM, Muir WW. Arrhythmogenicity of dopamine, dobutamine, and epinephrine in thiamylalhalothane anesthetized dogs. Am J Vet Res. 1983;44:2341–3.
12. Parratt JR, Wainwright CL, Fagbemi O. Effect of dopexamine hydrochloride in the early stages of experimental myocardial infarction and comparison with dopamine and dobutamine. Am J Cardiol. 1988;62:18C–23C.
13. Kirlin PC, Pitt B, Lucchesi BR. Comparative effects of prenalterol and dobutamine in a canine model of acute ischemic heart failure. J Cardiovasc Pharmacol. 1981;3:896–905.
14. Ozaki T, Uematsu T, Nagashima S, et al. Effects of DPI 201-106, a novel cardiotonic agent, on hemodynamics, cardiac electrophysiology and arrhythmias induced by programmed ventricular stimulation in dogs with subacute myocardial infarction: a comparative study with dobutamine. Arch Pharmacol. 1991;344:478–87.
15. Vanegas DI, de Montenegro J, de Perez C, Orjuela A. Dobutamine use for arrhythmia induction during electrical programmed heart stimulation. Revista Colombiana de Cardiologia. 2006;12:479–83.
16. Leier CV, Webel J, Bush CA. The cardiovascular effects of the continuous infusion of dobutamine in patients with severe cardiac failure. Circulation. 1977;56:468–72.
17. Leier CV, Heban PT, Huss P, et al. Comparative systemic and regional hemodynamic effects of dopamine and dobutamine in patients with cardiomyopathic heart failure. Circulation. 1978;58:466–75.

18. David S, Zaks JM. Arrhythmias associated with intermittent outpatient dobutamine infusion. Angiology. 1986;37:86–91.
19. Gillespie TA, Ambos HD, Sobel BE, et al. Effects of dobutamine in patients with acute myocardial infarction. Am J Cardiol. 1977;139:588–94.
20. Berthe C, Pierard LA, Hiernaux M, et al. Predicting the extent and location of coronary artery disease in acute myocardial infarction by echocardiography during dobutamine infusion. Am J Cardiol. 1986;158:1167–72.
21. Guimon J-G, Matuschak GM, Meyers F, et al. Augmentation of cardiac function in end-stage heart failure by combined use of dobutamine and amrinone. Chest. 1986;90:302–4.
22. Dies F, Krell MJ, Whitlow P. Intermittent dobutamine in ambulatory outpatients with chronic cardiac failure. Circulation. 1986;74:11–38.
23. Mertes H, Sawada SG, Ryan T, et al. Symptoms, adverse effects, and complications associated with dobutamine stress echocardiography: experience in 1118 patients. Circulation. 1993;88:15–9.

How to Induce Arrhythmias with Dopamine

9

Shemaila Saleem

9.1 Introduction

Dopamine is one of the most important neurotransmitters in the brain. It is a catecholamine derived from tyrosine and is the precursor of both noradrenaline and adrenaline (Fig. 9.1) [1].

S. Saleem (✉)
Al-Farabi Centre, Federal Medical and Dental College, Islamabad, Pakistan
e-mail:

G. Cismaru (ed.), *Arrhythmia Induction in the EP Lab*,
https://doi.org/10.1007/978-3-319-92729-9_9

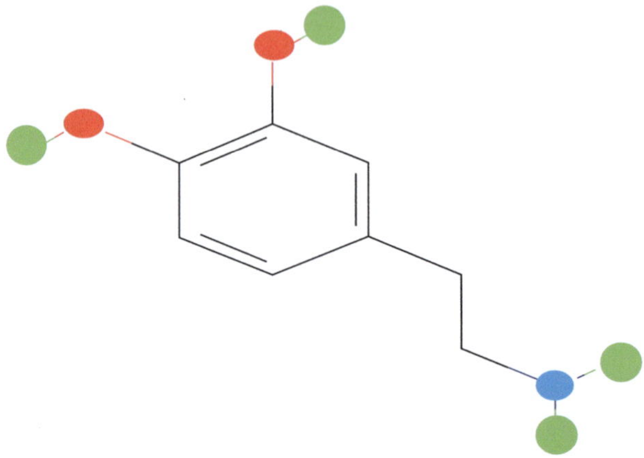

Fig. 9.1 Chemical structure of dopamine

Dopamine is a monoamine with positive inotropic activity. It binds to both α-1 and beta-1 adrenergic receptors. It increases the cardiac output by increasing the heart rate and contractility mediated through myocardial β-1 adrenergic receptors. Stimulation of α-1 adrenergic receptors on vascular smooth muscle leads to vasoconstriction and results in increased systemic vascular resistance. Dopaminergic receptors from the renal vasculature are also stimulated by dopamine which leads to dilation of renal blood vessels and increases glomerular filtration rate, renal blood flow, sodium excretion, and urine output [2].

Dopamine is commonly used for the correction of hemodynamic instability present in the shock due to a variety of conditions, including trauma, myocardial infarction, open-heart surgery, endotoxic septicemia, renal failure, and chronic cardiac decompensation in congestive failure [3].

Dopamine was reported to induce arrhythmias at usual doses employed for different types of shock. Atrial premature contractions, ventricular premature contractions, atrial flutter, non-sustained and sustained ventricular tachycardia were demonstrated at doses of 1–20 μg/kg/min [4].

9.2 Doses of Dopamine

At IV doses of 0.5–2 μg/kg/min, dopamine acts predominantly on dopaminergic receptors; at doses of 2–10 μg/kg/min, β1-adrenergic receptors are also stimulated by the drug (Fig. 9.2, Tables 9.1, 9.2 and 9.3). At higher therapeutic doses, the drug stimulates α-adrenergic receptors. The result of α-adrenergic receptors, β1-adrenergic, and dopaminergic stimulation produces the net effect of the drug, and the dose administered determines the main effects of dopamine. Cardiac stimulation and renal vascular dilation occur at lower doses, and vasoconstriction occurs at larger doses [5].

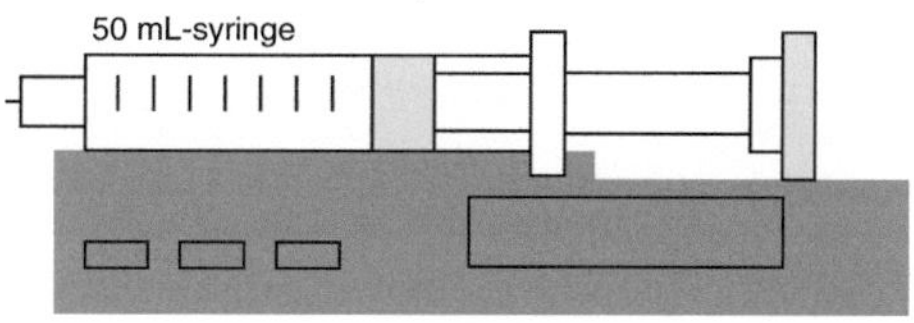

Fig. 9.2 Doses of dopamine

Table 9.1 Infusion preparation for Dopamine

Infusion preparation	
1.	The patient should not eat for at least 8 h before the electrophysiological study
2.	Patients are instructed to stop medication five times the half time prior to electrophysiological testing unless instructed otherwise
3.	An IV is started using aseptic technique
4.	An infusion pump is necessary for dopamine administration
5.	Dopamine is compatible with sodium chloride, dextrose, sodium lactate, and Ringer's lactate. Add 5 mL of 200 mg/5 mL dopamine vial into 50 cc syringe with 45 mL of NaCI 0.9% into the syringe
6.	Dopamine is started at a dose of 1 µg/kg/min and increased at 3-min interval by 5 µg/kg/min up to a maximum dose of 50 µg/kg/min
7.	Continuous monitoring of EKG, heart rate, and 1-min assessment of blood pressures and symptoms for each infusion stage

Table 9.2 Protocol for administration of Dopamine-dose increase per time unit

Time	Dose
Start	1 µg/kg/min
3 min	5 µg/kg/min
6 min	10 µg/kg/min
5 min	20 µg/kg/min
12 min	25 µg/kg/min
15 min	30 µg/kg/min
13 min	35 µg/kg/min
21 min	40 µg/kg/min
24 min	45 µg/kg/min
27 min	50 µg/kg/min

Table 9.3 Preparation of an infusion of Dopamine

Understanding the numbers:

You are ordered to give 20 μg/kg/min Dopamine to your 58-year-old, 70 kg patient for facilitate ventricular tachycardia induction by programmed stimulation

1. The Dopamine vial has 200 mg/5 mL
2. You should take 1 vial of Dopamine in the 50 mL syringe with 45 mL of serum (0.9% NaCl)
3. This gives a concentration of 200 mg/50 mL meaning 4 mg/mL
4. 20 μg x 70 kg=1400 μg should be injected, thus 1.4 mg equivalent of 0.35 mL/min
5. We will start the perfusion rate: 0.35 x 60=21 mL/h

Table 9.4 Contraindications for Dopamine

Contraindications for dopamine
• Blood pressure > 200/100 mmHg
• Recent myocardial infarction
• Unstable ischemic heart disease
• LV outflow tract obstruction
• Severe aortic stenosis
• Raynaud syndrome
• Critical limb ischemia

Table 9.5 Side effects of Dopamine

Side effects of dopamine
• Necrosis-if extravasation
• Angina
• Vasoconstriction
• Dyspnea
• Nausea
• Vomiting
• Headache
• Anxiety
• Pheochromocytoma

9.3 Clinical Use

One of the side effects of dopamine is the induction of ventricular tachyarrhythmias (Tables 9.4 and 9.5). Therefore in controlled settings, dopamine can be used to induce ventricular tachycardia for either benign forms like outflow tract VT or in patients with old ischemia, scar-dependent VT [6].

Dopamine can also induce other kinds of cardiac arrhythmias. For example, in one published case, due to the immaturity of the cardiac conduction system, dopamine may contribute to induction of supraventricular arrhythmia during early infancy [3]. In another case, electrocardiogram showed first-degree atrioventricular block before dopamine infusion, but electrocardiogram turned to second-degree atrioventricular block (Mobitz type II) after dopamine was started at an infusion rate of 10 μg/kg/min. When the infusion was gradually decreased to 4 μg/kg/min, ECG tracings returned to first-degree atrioventricular block. It was therefore theorized that the documented second-degree atrioventricular block may have been exaggerated by dopamine [4]. In another case, a patient with controlled atrial fibrillation-flutter and mitral stenosis developed a marked increase in atrioventricular conduction with high-rate atrial fibrillation during dopamine infusion [7].

9.4 Electrophysiological Effects of Dopamine

9.4.1 Animal Studies

Dopamine is a vital endogenous catecholamine known to modulate various physiological functions [8]. The most important is its functions on peripheral cardiovascular control. Dopamine increases heart rate and circulating levels of norepinephrine as well as epinephrine [9]. In animals, dopamine shortens ventricular repolarization and increases conduction velocity in the AV node and ventricles. This effect might be explained by the endogenous norepinephrine secretion from adrenergic presynaptic junctions [10].

Day et al. observed a dose-related increase in heart rate and blood pressure during administration of dopamine at doses of 30 and 45 μg intracerebroventricularly to a group of normotensive cats. This initial exaggerated response was followed by bradycardia and hypotension [8].

In another study on the impact of dopamine on the cardiovascular system of anesthetized dogs, it was found that dopamine, in small doses (2–4 μg/kg), had a minimal or no cardiac effect but a trivial pressor-depressor effect. Intermediate doses (8–16 μg/kg) were noted to augment the cardiac contractility and heart rate with significant increases in the blood pressure. A marked increase in heart rate, arterial pressure, and contractility was observed at higher doses (32–64 μg/kg). Exploration of the depressor phase induced by dopamine showed that administration of atropine, dibenzyline, hexamethonium, dichloroisoproterenol, or antihistamines did not block the hypotensive effect produced by dopamine. Depressor action of dopamine was studied by administering dopamine in constant perfusion of the limbs, and it was theorized that the observed effects were probably due to the reflex or a central nervous system effect [11].

Research on the adult and fetal sheep found a dose-related increase in the fetal pressor response with no effect of the advancing gestation on the magnitude of this response. Doses of 50–200 μg/kg of dopamine produce tachycardia (30–120 bpm) and an increased pressor response in atropinized fetuses when compared with unatropinized fetuses. Therefore the cardiovascular system of the fetus responds to large doses of dopamine which suggests that the mast cells need to secrete great quantities of endogenous dopamine to significantly modify the cardiovascular function of the sheep fetus [12]. In dogs, dopamine increases the inotropic function with no influence on the heart rate. Propranolol abolishes this effect, whereas reserpine reduces the inotropic response. However atropine has no influence on the dopamine-induced inotropic effects [13].

Holmes et al. observed that in dogs' heart-lung preparation, dopamine exhibits marked inotropic and chronotropic effects. They further elaborated that in uncontrolled rate preparations, dopamine doses of 0.5–1 mg cause an 18% augmentation of the heart rate. Furthermore, dopamine dose of 100–500 μg was noted to amplify contractility of both ventricles, blood pressure, and cardiac output with a consistent attenuation of atrial pressures in controlled rate preparations [14].

Tisdale et al. found that dopamine had a biphasic effect on the duration of action potential. It also causes an upsurge in the automaticity in Purkinje fibers. Dopamine caused both atrial and ventricular tachyarrhythmias in animals [15].

9.4.2 Human Studies

Generally, dopamine receptors are activated by low doses of dopamine (0.5–2.0 μg/kg/min), which causes a decrease in peripheral vascular resistance and left ventricular afterload. With the increase in dose (2–4 μg/kg/min), there was noted activation of β1-adrenoceptors which, in turn, leads to increase in the cardiac output. A further increase in the dose (5 μg/kg/min) stimulates α-adrenoceptors resulting in vasoconstriction and amplified peripheral vascular resistance.

Council et al. showed that at doses ranging from 3 to 7.5 μg/kg/min, dopamine functions as a β-agonist, amplifying the heart rate and cardiac output. Dopamine, when compared with isoproterenol, was found to have less effective β-agonist effects [16]. In comparison with dobutamine, the inotropic effects of dopamine are modest [17]. Nowadays dopamine is considered to be a safe drug which can be used as a substitute for isoproterenol to manage bradycardias in which atropine administration is either futile or contraindicated. A dose- dependent escalation in pulmonary capillary wedge pressure demonstrates dopamine-induced vasoconstriction of pulmonary veins, and there is a spurious elevation of LV filling pressures.

Bischoff et al. studied the electrophysiologic effects of dopamine at doses of 3 and 6 μg/min in patients with and without various rhythm disturbances using right atrial pacing and His-Bundle electrography. The blood pressure was found to increase significantly. They demonstrated that electrophysiologic parameters of the sinus and atrium were not markedly altered. The AV-nodal conduction was not altered by the 6 μg dopamine. Improvement, however, was noticed in intranodal conduction by 3 μg dopamine administration. The intraventricular conduction time was also found to improve insignificantly, with no improvement seen in the His-Purkinje system aberrant conduction provoked by paired stimulation [18]. Human data have revealed dose-related sinus tachycardia, with some cases of ventricular arrhythmias.

9.4.2.1 Dopamine for Supraventricular Arrhythmias Induction

Depending on the dosage, dopamine acts in three distinct ways. Low doses of dopamine (1–2 μg/kg/min) cause vasodilatation. Intermediate doses (2–10 μg/kg/min) stimulate β-adrenergic function and increase the cardiac output. A dose of

10 μg/kg/min results in a potent vasoconstrictor effect. Due to the strong vasoconstrictor function of dopamine, one should be extremely cautious to avoid the direct leakage of infusions into the tissue [19].

Dopamine is an effective inotropic agent and may also induce dose-related sinus tachycardia. Shahar et al. describe a case of an infant with heart failure in whom infusion of dopamine resulted in an episode of paroxysmal supraventricular tachycardia. During early period of infancy, immature or unstable cardiac conduction system may lead to the induction of arrhythmias of supraventricular origin [3]. Research has revealed that in animals dopamine may induce atrial or ventricular arrhythmias.

9.4.2.2 Dopamine for Atrial Fibrillation Induction

Dopamine can produce atrial fibrillation in individuals with acute decompensated cardiac failure, and if administered after open-heart surgery in hypotensive patients, it causes an increased risk of postoperative atrial fibrillation [20]. Dopamine increases cAMP levels by activating β1-adrenergic receptors. It shortens atrial action potential duration and atrial effective refractory period and also increases focal ectopic automaticity resulting from delayed or early after depolarizations in the pulmonary veins. Furthermore, atrioventricular nodal conduction is also increased. Also, patients with atrial fibrillation, if infused with dopamine, may experience increase in the ventricular rate [21, 22].

9.4.2.3 Dopamine for Tachyarrhythmia Induction

Dopamine can also induce asymptomatic ventricular ectopic activity. At any particular cardiac index, dopamine induces marked increase in the heart rate or incidence of ventricular premature beats [23]. Bednarski et al. explained that dopamine is capable of producing cardiac arrhythmias in vagotomized as well as non-vagotomized anesthetized dogs. Moreover bilateral vagotomy can lead to a decreased requirement of dopamine dosage to induce cardiac arrhythmias. He also stated that atrial arrhythmias can be induced by dopamine dosages lesser than those essential for producing ventricular arrhythmias [10].

9.4.2.4 Dopamine for Bradycardia Induction

Tsompanidou et al., in their experiments on postsurgery dogs anesthetized with isoflurane, observed the development of sudden and profound bradycardia on administration of dopamine infusion. Bradycardia is believed to be triggered by the inhibitory reflex called Bezold-Jarisch reflex with characteristic bradycardia and hypotension [24].

Perez-Olea et al. observed that the early hypotensive phase and bradycardia, as a consequence of dopamine infusion, occurs as a result of vagal reflex. Vagotomy prevented dopamine-induced hypotension as well as bradycardia [25].

In rats pretreated with atenolol, the effect of dopamine agonists on bradycardia induced by vagal stimulation was studied. The dopamine agonists decreased the vagal-induced bradycardia significantly, but there was no effect on acetylcholine-induced bradycardia [26].

9.4.2.5 Dopamine for Atrioventricular Conduction

Gelfman et al. reported an individual with atrial fibrillation-flutter who experienced a significant increase in atrioventricular conduction during the administration of low-dose dopamine. In humans this response has not been previously addressed [7].

9.4.2.6 Dopamine for Ventricular Tachycardia Induction

Dopamine-associated ventricular tachycardias rarely occur. Orth et al. have stated in their research that dopamine caused the induction of ventricular tachycardia in four out of five experimental dogs anesthetized by cyclopropane [27].

Loeb and colleagues have also observed the development of ventricular tachycardia in one of their patients immediately after infusion of dopamine. The patient was suffering from shock due to acute myocardial infarction complicated by atrioventricular block. It was noticed that soon after the infusion was discontinued, the tachycardia disappeared [28].

References

1. McGrath BP, Wang XQ. Dopamine: clinical applications iii. Cardiovascular. Aust Prescr. 1994;17:44–5.
2. Wick M. Therapeutic effect of dopamine infusion on human malignant melanoma. Cancer Treat Rep. 1982;66:1657–9.
3. Shahar E, Lotan D, Barzilay Z. Dopamine-induced paroxysmal supraventricular tachycardia in an infant. Clin Pediatr. 1981;20:541–3.
4. Kitamura T, Ohno N, Bougaki M, Saito Y, Orii R, Yamada Y. Second-degree atrioventricular block (Mobitz type II) probably exaggerated by dopamine during anesthetic management for a patient with ochronosis: a case report. Masui. 2008;57:1513–6.
5. Ozaki T, Uematsu T, Nagashima S, Nishimoto M, Nakashima M. Effects of DPI 201-106, a novel cardiotonic agent, on hemodynamics, cardiac electrophysiology and arrhythmias induced by programmed ventricular stimulation in dogs with subacute myocardial infarction: a comparative study with dobutamine. Naunyn Schmiedebergs Arch Pharmacol. 1991;344:478–87.
6. Tisdale JE, Patel R, Webb CR, Borzak S, Zarowitz BJ. Electrophysiologic and proarrhythmic effects of intravenous inotropic agents. Prog Cardiovasc Dis. 1995;38:167–80.
7. Gelfman DM, Ornato JP, Gonzalez ER. Dopamine- induced increase in atrioventricular conduction in atrial fibrillation-flutter. Clin Cardiol. 1987;10:671–3.
8. Day M, Roach A. Cardiovascular effects of dopamine after central administration into conscious cats. Br J Pharmacol. 1976;58:505–15.
9. Ziegler M, Kennedy B, Holland O, Murphy D, Lake C. The effects of dopamine agonists on human cardiovascular and sympathetic nervous systems. Int J Clin Pharmacol Ther Toxicol. 1985;23:175–9.
10. Harris W, Van Petten G. The effects of dopamine on blood pressure and heart rate of the unanesthetized fetal lamb. Am J Obstet Gynecol. 1978;130:211–5.
11. McDonald RH, Goldberg LI. Analysis of the cardiovascular effects of dopamine in the dog. J Pharmacol Exp Ther. 1963;140:60–6.
12. Gilbert J, Lange G, Polevoy I, Brooks CM. Effects of vasoconstrictor agents on cardiac irritability. J Pharmacol Exp Ther. 1958;123:9–15.
13. Alousi A. Cardiovascular effects of dopamine in the normal and failing heart of experimental animals. Recent Adv Stud Cardiac Struct Metab. 1975;6:291–300.
14. Holmes JC, Fowler NO. Direct cardiac effects of dopamine. Circ Res. 1962;10:68–72.

15. Tisdale JE, Patel RV, Webb CR, Borzak S, Zarowitz BJ. Proarrhythmic effects of intravenous vasopressors. Ann Pharmacother. 1995;29:269–81.
16. European Resuscitation Council. Part 6: advanced cardiovascular life support. Section 6: pharmacology II: agents to optimize cardiac output and blood pressure. European Resuscitation Council. Resuscitation. 2000;46:169–84.
17. Imanaga I, Kaneda T, Miyakawa N. Comparison of the effects of dobutamine with dopamine and halothane anesthetized dogs. American journal of isoproterenol on inotropism and chronotropism in the mammalian heart (author's transl). Nihon Yakurigaku Zasshi. 1979;75(2):147–57.
18. Bischoff K, Hager W, Bödeker K. Electrophysiologic effects of dopamine on the cardiac conduction system (author's transl). Z Kardiol. 1981;70(9):643–51.
19. Stetson JB, Reading GP. Avoidance of vascular complications associated with the use of dopamine. Can Anaesth Soc J. 1977;24(6):727–33.
20. Argalious M, Motta P, Khandwala F, Samuel S, Koch CG, Gillinov AM, et al. "Renal dose" dopamine is associated with the risk of new-onset atrial fibrillation after cardiac surgery. Crit Care Med. 2005;33(6):1327–32.
21. Sampson KJ, Terrenoire C, Cervantes DO, Kaba RA, Peters NS, Kass RS. Adrenergic regulation of a key cardiac potassium channel can contribute to atrial fibrillation: evidence from an IKs transgenic mouse. J Physiol. 2008;586(2):627–37.
22. Patterson E, Yu X, Huang S, Garrett M, Kem DC. Suppression of autonomic-mediated triggered firing in pulmonary vein preparations, 24 hours postcoronary artery ligation in dogs. J Cardiovasc Electrophysiol. 2006;17(7):763–70.
23. Bednarski RM, Muir W. Arrhythmogenicity of dopamine, dobutamine, and epinephrine in thiamylal-halothane anesthetized dogs. Am J Vet Res. 1983;44(12):2341–3.
24. Tsompanidou PP, Kazakos GM, Anagnostou TL. Dopamine-induced bradycardia in two dogs under isoflurane anaesthesia. J Small Anim Pract. 2013;54(12):672–4.
25. Perez-Olea J, Quevedo M, Silva R. Enhancement of blood pressure response to dopamine by angiotensin II. Hypertension. 1981;3(6 Pt 2):II-138–41.
26. Roquebert J, Moran A, Demichel P, Sauvage M-F. Pharmacological characterization of dopamine receptors in parasympathetic innervation of rat heart. Eur J Pharmacol. 1991;200(1):59–63.
27. Orth OS, Stutzman JW, MEEK WJ. Relationship of chemical structure of sympathomimetic amines to ventricular tachycardia during cyclopropane anesthesia. J Pharmacol Exp Ther. 1944;81(2):197–202.
28. Loeb HS, Winslow EB, Rahimtoola SH, Rosen KM, Gunnar RM. Acute hemodynamic effects of dopamine in patients with shock. Circulation. 1971;44(2):163–73.

Adrenaline Versus Atropine for AVNRT Induction

10

Jennifer Asamoah

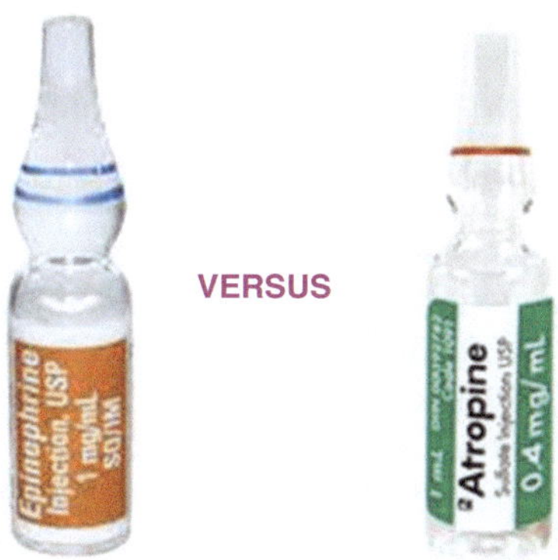

10.1 Introduction

AVNRT is the most frequent form of paroxysmal supraventricular tachycardia. In French the term is TRIN (tachycardie par reentre intra-nodale) and in English AVNRT (atrioventricular node reentrant tachycardia). It can be induced in the EP lab by atrial or ventricular stimulation in the basal state or after drug administration. The drug of choice is isoprenaline, but in some European countries, the lack of the drug produced efforts to search for another options. There is no study that compares efficacy of atropine versus adrenaline for AVNRT induction in the EP lab. In our study, we compared the two drugs in terms of efficacy and side effects.

J. Asamoah (✉)
New York University School of Medicine,
550 1st Avenue, New York, NY, USA

G. Cismaru (ed.), *Arrhythmia Induction in the EP Lab*,
https://doi.org/10.1007/978-3-319-92729-9_10

10.2 Hypothesis

In order to compare atropine with adrenaline, we took patients with AVNRT that received atropine and compared them with patients that received adrenaline, without corresponding placebo matching per intervention group. The investigators hypothesized that adrenaline would be a better and more convenient for the induction of arrhythmias since adrenaline is administered as an intravenous infusion over several minutes, in comparison to atropine, which is administered as a bolus. We demonstrated our hypothesis by statistical analysis.

10.3 Materials and Method

10.3.1 Study Population and Methodology

This study was a retrospective, observational study which analyzed 216 consecutive patients who underwent AVNRT ablation and received either adrenaline (49 patients) or atropine (167 patients) for arrhythmia induction. It was ensured that informed consent was obtained before EP study for every subject included in the analysis. Each subject had a history of AVNRT with an ECG showing retrograde P wave in inferior leads: D2, D3, avF, or rsR' in lead V1. Patients that required both atropine and adrenaline for AVNRT induction were excluded from the study. Further exclusion criteria included the presence of prostatic adenoma or glaucoma with closed angle when chosen to have received atropine. Due to retrospective nature of the study, the authors note the limitations in statistical analysis due to the differing numbers of subjects that received adrenaline and atropine.

10.3.2 Electrophysiological Study

Electrophysiological study was performed without general anesthesia or sedation. The only anesthetic used was local anesthesia (lidocaine 1%) that was administered subcutaneously. All patients were noted to be in sinus rhythm at the beginning of the electrophysiological study. Catheters were introduced via femoral, subclavian or jugular venous access sites. The right side was preferred over the left side. For most of the procedures, four catheters were used according to our department's protocol: one in the superior right atrium, one in the region of the His bundle, one inside the coronary sinus, and one at the apex of the right ventricle.

We performed atrial and ventricular stimulation to measure the refractory period of the atrium, ventricle, and AV node. We excluded the presence of an accessory pathway by atrial and ventricular stimulation and by demonstrating absence of retrograde conduction to the atrium or when present decremental conduction with the first atrial depolarization at the level of the His bundle. After arrhythmia

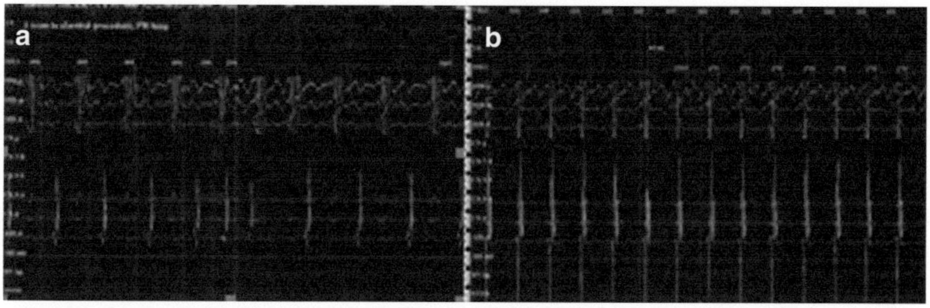

Fig. 10.1 Atrial stimulation for AVNRT induction. Surface and intracardiac derivations can be seen (from up to down): the first three derivations in green are surface ECG; the following in red is high right atrium (HRA); in blue the His derivations, proximal Hisp, medium Hism, and distal His, Hisd; in yellow the coronary sinus, distal and proximal; in white the ablation catheter. The tachycardia cycle length is 270 ms with the ventricular electrogram superimposed with the atrial electrogram which is suggestive of typical AVNRT. (**a**) Before adrenaline infusion, atrial extrastimuli could not induce AVNRT; (**b**) after adrenaline infusion, programmed atrial stimulation induces AVNRT

induction, we used atrial and ventricular entrainment to confirm AVNRT by measuring the post-pacing interval with orthodromic reentrant tachycardia and atrial tachycardia excluded. We stimulated the atrium with up to three extrastimuli at the level of high right atrium and coronary sinus on a 600 and 400 ms imposed rhythm. If the clinical arrhythmia could not be induced with extrastimuli, then burst atrial or ventricular pacing was performed. If clinical arrhythmias could not be induced, a bolus of either atropine 1 mg or infused adrenaline to facilitate arrhythmia induction. The choice between the two drugs was left at the discretion of the interventionist. Adrenaline was infused at an initial dose of 0.05 μg/kg/min and increased by 0.05 μg/kg/min until the heart rate increased by 50% to at least 100 bpm but not more than 150 bpm (cycle length of 400 ms). The dose of adrenaline that permitted increase of the heart rate was noted to be 0.1–0.3 μg/kg/min (Fig. 10.1). If atropine was injected, a bolus of 1–2 mg was used. The medium dose of atropine used was 1.5 mg with a minimum of 1 mg and a maximum of 3 mg. If the clinical arrhythmia cannot be induced with one of the two drugs, then the other one was added, but patients that received both atropine and adrenaline were excluded from the study.

10.3.3 Statistical Analysis

Statistical analysis was done using SPSS version 22. For descriptive statistics we used mean and standard deviation in case of normally distributed values. For continuous variables without normal distribution, we used median and interquartile range. For comparison between atropine and adrenaline, we used either the chi-square test or the Fisher's exact test. For comparison of normally distributed variables, we used the Student's *T*-test and, in case of abnormal distribution, the Mann-Whitney test.

10.4 Results

10.4.1 Patients' Characteristics

A total of 49 patients received adrenaline, and 167 patients received atropine for AVNRT induction. Due to the retrospective nature of this study, no baseline population matching between the two groups could be done. The median dose of adrenaline was 0.2 μg/kg/min, and the median dose of atropine was 1.5 mg/patient.

The mean age of the adrenaline group was 50.4 years (range 17–68), and the mean age of the atropine group was 46.3 years (range 15–82) with a significant difference between groups ($p = 0.0002$) (Fig. 10.2). In the atropine group, 63.3% were females (33 of 49), and in the adrenaline group, 58.9% (99/167) were females. We compared both groups in terms of sex distribution, and we found a nonsignificant difference with a p value of 0.586 (Fig. 10.3, Table 10.1).

The group of patients that was given adrenaline was able to achieve their primary endpoint (AVNRT induction) in 40 out of 49 patients (81.5%), while the atropine group achieved the endpoint in 143 out of 167 patients (85.7%). Although the percentage is higher in the atropine group, the statistical chi-square test showed a nonsignificant difference between the two groups with a *p* value of 0.494 (Figs. 10.4 and 10.5, Table 10.2).

10.4.2 Side Effects of Atropine and Adrenaline

The adrenaline infusion had to be stopped in 2 of the 49 patients (4%) because of they developed headache at blood pressure values of more than 180/110 mmHg, for infusion rates of more than 0.2 μg/kg/min. Blood pressure was subsequently lowered

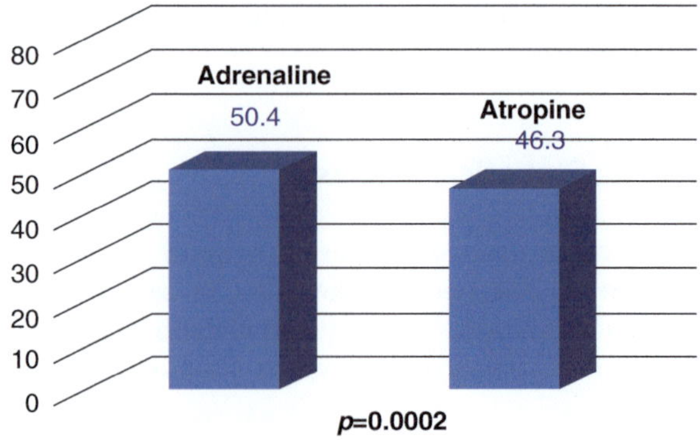

Fig. 10.2 Comparison of age between the two groups found a significant *p* value of 0.0002

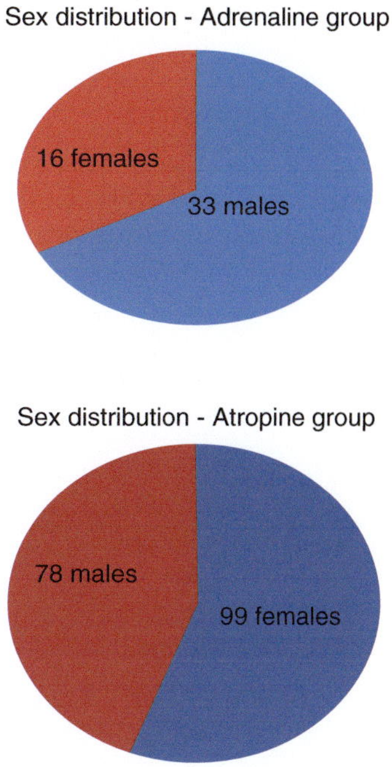

Fig. 10.3 Sex distribution between the two groups shows a nonsignificant difference with a *p* value of 0.586

Table 10.1 General characteristics of patients that received Atropine or Adrenaline

	Adrenaline	Atropine	*P*
Number of patients	49	168	
Mean age	50.4 (17–68)	46.3 (15–82)	0.0002
Females	31 (63.3%)	99 (58.9%)	0.586
Males	18 (36.7)	69 (41.1)	0.586

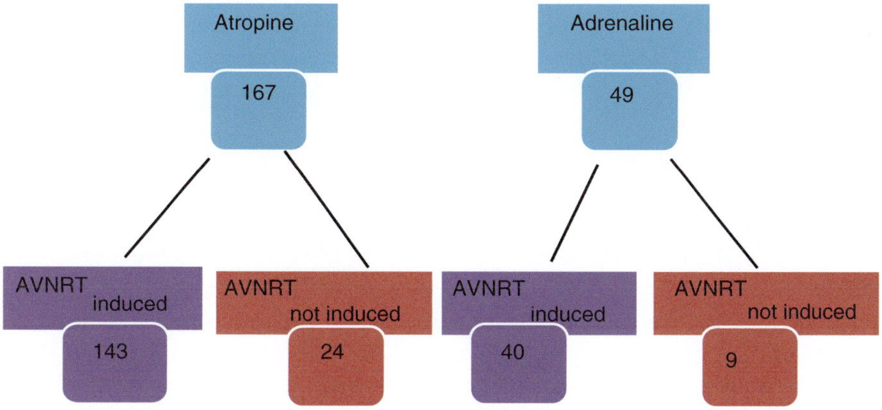

Fig. 10.4 AVNRT induction with atropine and with adrenaline

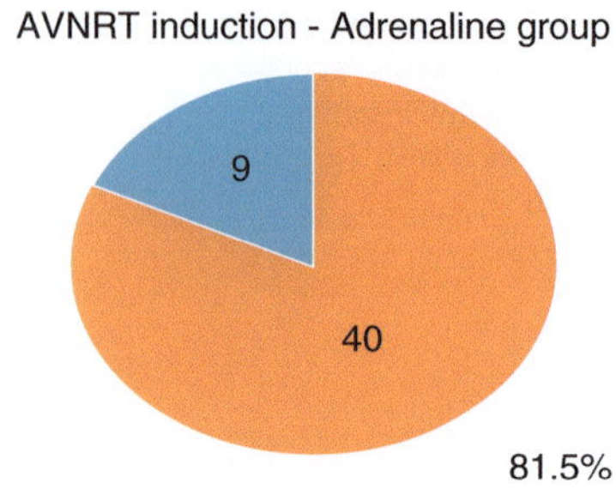

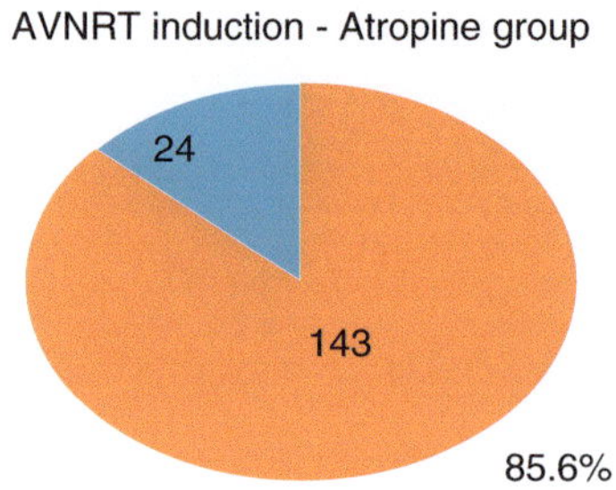

Fig. 10.5 AVNRT induction with atropine and adrenaline. Chi-square test shows a nonsignificant difference between the two groups with a *p* value of 0.494

Table 10.2 The chi-square test for comparison between the 2 groups shows a non-significant *p* value of more than 0.4

		Levene's test for equality of variances		*t*-test for Equality of Means						
									95% confidence interval of the difference	
		F	Sig.	*t*	df	Sig. (2-tailed)	Mean difference	Std. error difference	Lower	Upper
ATROPINE_ vs._ ADRENALINE	Equal variances assumed	1.752	0.187	0.681	214	0.496	0.03996	0.05866	−0.07567	0.15559
	Equal variances not assumed			0.643	72.309	0.522	0.03996	0.06217	−0.08396	0.16388

with intravenous enalapril. Tremors also appeared at doses >0.1 μg/kg/min. Because tremors are well tolerated by the patients, the infusion was not stopped for this particular side effect.

In the atropine group, the most frequent side effect was dry mouth, which was well tolerated by the patients during all the electrophysiological studies.

During the adrenaline infusion, atrial premature beats and ventricular premature beats were rarely seen at doses >0.2 μg/kg/min. After atropine administration, no patients presented with atrial or ventricular premature beats. No patients in the atropine group had closed-angle glaucoma. None of the patients with prostatic adenoma that received atropine had acute retention of urine.

10.5 Discussion

10.5.1 Results Interpretation

Our study demonstrated that both atropine and adrenaline had significant capabilities to induce AVNRT: 81.6% for adrenaline and 85.7% for atropine. The difference of 4% between the two drugs is nonsignificant, despite atropine having a higher percentage. However, the investigators understand the limitations of the study and understand that a higher number of patients in the adrenaline group would be necessary to confirm differences between the two groups.

Adrenaline was generally well tolerated, with the investigators noting that the infusion had to be stopped in only 2 out of the 49 cases where it was administered. None of the patients from the atropine group had significant side effects.

We believe that both atropine and adrenaline can be used in the EP lab for arrhythmia induction when isoprenaline, being the current gold standard, is not readily available. However, the main disadvantage of adrenaline is the necessity of an infusion and a longer time to achieve its full effect compared with atropine, which is administered as a bolus with a rapid onset of action. As previously reiterated, in patients with closed-angle glaucoma and prostatic adenoma, adrenaline may be preferred over atropine.

10.5.2 Prior Studies with Atropine

The first study that demonstrated the utility of drugs for arrhythmia induction in the EP lab was that of Akhtar et al. [1] published in 1975, which demonstrated that atropine was capable of increasing patient susceptibility to paroxysmal supraventricular tachycardia during programmed stimulation. Hariman et al. [2] showed that atropine could be used for atrioventricular node reentrant tachycardia induction because it facilitates retrograde ventriculoatrial conduction. Wu et al. [3] showed increased conduction through the antegrade and retrograde conduction after atropine. Patients that were inducible after atropine showed a similar refractory period of both fast and slow nodal pathways after atropine. In the study of Kim et al. [4], atropine added to programmed stimulation facilitated reentry through an accessory pathway and paroxysmal supraventricular tachycardia in 2 of 18 patients with asymptomatic WPW syndrome.

10.5.3 Prior Studies with Adrenaline

The effects of adrenaline in patients with supraventricular tachycardias were previously studied, but at present, there is no data showing the efficacy of this drug for PSVT induction. Increase in plasma concentration of adrenaline leads to a decrease in atrial refractory period, a decrease in refractory period of the AV node, and an increase in the conduction velocity through the AV node.

In the study by Cismaru et al. [5], an increase in the plasma concentration of adrenaline can lead to PSVT induction in patients that are non-inducible in the basal state.

10.5.4 Prior Studies with Isoprenaline

Hariman et al. [2] were the first to show that isoprenaline was effective for PSVT induction for cases that were non-inducible in the basal state or after atropine injection. They documented a 79-year-old patient that received isoprenaline infusion with PSVT subsequently induced. Toda et al. [6] demonstrated a 50% sensitivity for isoprenaline in the induction of paroxysmal supraventricular tachycardia presented on exertion. Huycke et al. [7] noted a sensitivity of 67% and a specificity of 100% for PSVT induction in a case series of 20 patients with dual nodal pathway, without any arrhythmia induced in the basal state. In the same year, Brembilla-Perot et al. [8] published their results obtained in Nancy Center, France, and demonstrated a sensitivity of 90% for PSVT induced on exertion (28 out of 31 patients). In the control group (of 37 patients without PSVT), after isoprenaline infusion, no arrhythmia was induced which further permitted to calculate a 100% specificity for isoprenaline for PSVT induction. Stellbrink et al. [9] demonstrated induction of AVNRT via isoprenaline infusion in 93% of cases. Katz et al. [10] reported a 46% induction rate of PSVT in patients with ventricular preexcitation. Oral et al. [11] demonstrated a sensitivity of 80% and a specificity of 95% for atrial fibrillation induction after 20 μg/min of isoprenaline was administered to patients with paroxysmal, persistent or long-term persistent atrial fibrillation.

10.5.5 Limitations

This study has two important limitations: Firstly, the investigators acknowledge that this is a non-randomized study, with information obtained in a retrospective manner. There were therefore significant differences in baseline characteristics with large differences in the population for both groups analyzed.

Second, in patients that received atropine for arrhythmia induction, the absence of a slow pathway at the end of procedure could be questionable. Lin et al. showed that after atropine injection, the refractory period of the slow pathway is close to the refractory period of the fast pathway and conduction jumps could be absent in this case. At the end of the procedure, the absence of the conduction jump may be the effect of high atropine concentration in the blood of the patient and not the sign of an effective slow pathway ablation [12].

10.6 Conclusion

Adrenaline and atropine demonstrated similar efficacies for AVNRT induction in the EP laboratory. Both may be used when isoprenaline is not available, as is seen in some Eastern and Western European countries. The study reports an induction rate of 81.5% for adrenaline and 85.6% for atropine with no significant differences between each drug. However, the investigators recognize the limitations of the retrospective manner of data collection, particularly in the differences among baseline characteristics between compared groups and the number of subjects per groups.

References

1. Akhtar M, Damato AN, Batsford WP, Caracta AR, Ruskin JN, Weisfogel GM, et al. Induction of atrioventricular nodal reentrant tachycardia after atropine. Report of five cases. Am J Cardiol. 1975;36:286–91.
2. Hariman RJ, Gomes JA, El-Sherif N. Catecholamine- dependent atrioventricular nodal reentrant tachycardia. Circulation. 1983;67:681–6.
3. Wu D. A-V nodal reentry. Pacing Clin Electrophysiol. 1983;6:1190–200.
4. Kim YH, Leitch JW, Klein GJ, Yee R, Leather RA. Effect of atropine and isoproterenol on tachycardia induction in asymptomatic patients with Wolff- Parkinson-White electrocardiographic pattern. Can J Cardiol. 1992;8:716–20.
5. Cismaru G, Rosu R, Muresan L, Puiu M, Andronache M, Hengan E, et al. The value of adrenaline in the induction of supraventricular tachycardia in the EP lab. Europace. 2014;16:1634–8.
6. Toda I, Kawahara T, Murakawa Y, Nozaki A, Kawakubo K, Inoue H, et al. Electrophysiological study of young patients with exercise related paroxysmal of palpitation: role of atropine and isoprenaline for initiation of supraventricular tachycardia. Br Heart J. 1989;61:268–73.
7. Huycke EC, Lai WT, Nguyen NX, Keung EC, Sung RJ. Role of intravenous isoproterenol in the electrophysiological induction of atrioventricular node reentrant tachycardia in patients with dual atrioventricular node pathways. Am J Cardiol. 1989;64(18):1131–7.
8. Brembilla- Perrot B, Terrier de la Chaise A, Pichene M, Aliot E, Cherrier F, Pernot C. Isoprenaline as an aid to the induction of catecholamine dependent supraventricular tachycardia during programmed stimulation. Br Heart J. 1989;61:348–55.
9. Stellbrink C, Diem B, Schauerte P, Brehmer K, Schuett H, Hanrath P. Differential effects of atropine and isoproterenol on inducibility of atrioventricular nodal reentrant tachycardia. J Interv Card Electrophysiol. 2001;5:463–9.
10. Katz DF, Hegeman TW, Aleong R. Use of isoproterenol for assessment of anterograde accessory pathway conduction properties in sedated asymptomatic patients with ventricular preexcitation. J Innov Cardiac Rhythm Manage. 2013;4:1428–33.
11. Oral H, Crawford T, Frederick M, Gadeela N, Wimmer A, Dey S, et al. Inducibility of paroxysmal atrial fibrillation by isoproterenol and its relation to the mode of onset of atrial fibrillation. J Cardiovasc Electrophysiol. 2008;19:466–70.
12. Katritsis DG, Camm AJ. Classification and differential diagnosis of atrio-ventricular nodal reentrant tachycardia. Europace. 2006;8:29–36.

11 Adrenaline Versus Isoprenaline for Arrhythmia Induction

Andrei Cismaru

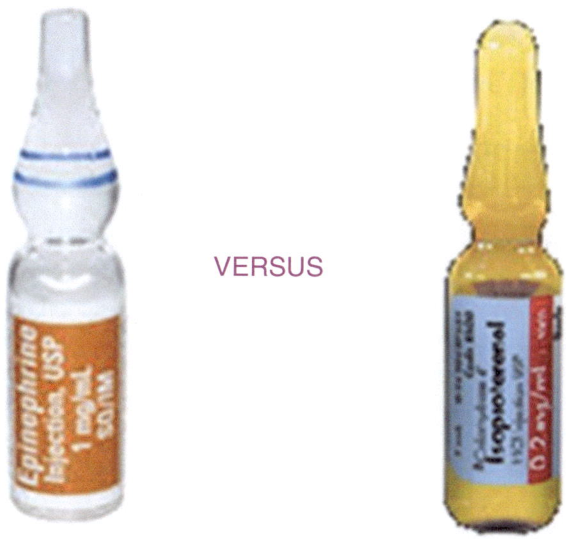

11.1 Adrenaline Versus Isoprenaline for AVNRT Induction

11.1.1 Introduction

AVNRT can be induced in the EP lab by atrial or ventricular stimulation in the basal state or after drug administration. The drug of choice is isoprenaline, but in some European countries, the lack of the drug demands efforts to search for other options.

A. Cismaru (✉)
Institut Gustave Roussy, Villejuif, France
e-mail: cosmin.cismaru@umfcluj.ro, cosmin-andrei.cismaru@gustaveroussy.fr

G. Cismaru (ed.), *Arrhythmia Induction in the EP Lab*,
https://doi.org/10.1007/978-3-319-92729-9_11

To date, there is no study that compares efficacy of isoprenaline versus adrenaline for AVNRT induction in the EP lab. In our study we compared the two drugs in terms of efficacy and side effects. The results shows that both isoprenaline and adrenaline have a high sensibility for AVNRT induction: 81.6% for adrenaline and 90% for isoprenaline. The difference of 8.4% between the two drugs is nonsignificant, although it looks that isoprenaline would have a better profile. A higher number of patients are necessary to confirm the difference between the two drugs.

11.1.2 Hypothesis

In order to compare isoprenaline with adrenaline, we included all patients with AVNRT that received isoprenaline and all patients that received adrenaline. No match was done between the two groups because the number of patients in the two groups was low.

11.1.3 Materials and Method

11.1.3.1 Study Population

We studied a consecutive group of patients with AVNRT ablation and no arrhythmia induced in the basal state. For arrhythmia induction we infused adrenaline in 49 patients and isoprenaline in 20 patients. Every patient had a history of AVNRT with an ECG showing retrograde P wave in inferior leads: D2, D3, avF, or rsR′ in lead V1. The study was retrospective and observational without any action on the study group. We did not select the patients; we did not perform a "match" of the patients in function of the age or sex; we just included all the patients that required isoprenaline or adrenaline treatment in order to obtain data from "real-world" patients.

The isoprenaline group was composed of 20 patients, 18 female and 2 male patients. The adrenaline group was composed of 16 female and 33 male patients.

11.1.3.2 Electrophysiological Study

Electrophysiological study was performed without general anesthesia or sedation. The only anesthetic used was 1% lidocaine subcutaneously at the site of catheter insertion. All patients were in sinus rhythm at the beginning of the electrophysiological study. Catheters were inserted inside the heart chambers using the femoral, subclavian, or jugular vein. The right side was preferred over the left side. For most of the procedures, four catheters were used according to our department protocol: one in the superior right atrium, one in the region of the His bundle, one inside the coronary sinus, and one at the apex of the right ventricle.

We performed atrial and ventricular stimulation to measure the refractory period of the atrium, ventricle, and AV node. We excluded the presence of an accessory pathway by atrial and ventricular stimulation and by demonstrating absence of retrograde conduction to the atrium or when present decremental conduction with the first atrial depolarization at the level of the His bundle. After arrhythmia induction we used atrial and ventricular entrainment to confirm AVNRT by measuring the post-pacing interval, and we excluded orthodromic reentrant tachycardia or atrial tachycardia. We stimulated the atrium with up to three extrastimuli at the level of high right atrium and coronary sinus on a 600 and 400 imposed rhythm. If the clinical arrhythmia could not be induced with extrastimuli, then burst atrial or ventricular pacing was performed.

When clinical arrhythmia could not be induced, we infused 1 μg/kg/min isoprenaline or infused 0.1 μg/kg/min adrenaline to facilitate arrhythmia induction. The choice between the two drugs was left at the discretion of the interventionist. When adrenaline was infused, an initial dose of 0.05 μg/kg/min was used and increased by 0.05 μg/kg/min every 5 min until the heart rate increased by 50% to at least 100 bpm but not more than 150 bpm (cycle length of 400 ms). The dose of adrenaline that permitted increase of the heart rate was 0.1–0.3 μg/kg/min (Fig. 11.1). When isoprenaline was infused, it was started at a dose of 0.5 μg/kg/min and then increased with 0.5 μg/kg/min every 3 min until the heart rate increased by 50% to at least 100 bpm but not more than 150 bpm. The medium dose of isoprenaline was 5 μg/kg/min. No patients received both drugs in association.

11.1.3.3 Statistical Analysis

Statistical analysis was performed using the 22 SPSS version. For descriptive statistics we used mean and standard deviation in case of normally distributed values. For continuous variables without normal distribution, we used median and interquartile range. For comparison between isoprenaline and adrenaline, we used the chi square test and the Fisher test when appropriate. For comparison of normally distributed variables, we used the student test and in case of abnormal distribution, the Mann-Whitney U test.

11.1.4 Results

In the adrenaline group, AVNRT was induced in 40 of 49 patients (81.5%) and in the isoprenaline group 18 of 40 patients (90%). Although the percentage was higher in the isoprenaline group, the statistical test showed a nonsignificant difference between the two groups with a p value of 0.396 (Table 11.1).

11.1.4.1 Side Effects of Isoprenaline and Adrenaline

Adrenaline infusion had to be stopped in 2 of the 49 patients (4%) because of headache at blood pressure values of more than 180/110 mmHg, for infusion rates of more than 0.2 μg/kg/min. Blood pressure was lowered with intravenous enalaprilat. Tremor also appeared at doses >0.1 μg/kg/min. Because tremor was well tolerated by the patients, infusion was not stopped for this side effect.

In the isoprenaline group, the most frequent side effects were agitation and tremor, but they were well tolerated by the patients during the electrophysiological study.

During adrenaline infusion atrial premature beats and ventricular premature beats were rarely seen at doses >0.2 μg/kg/min.

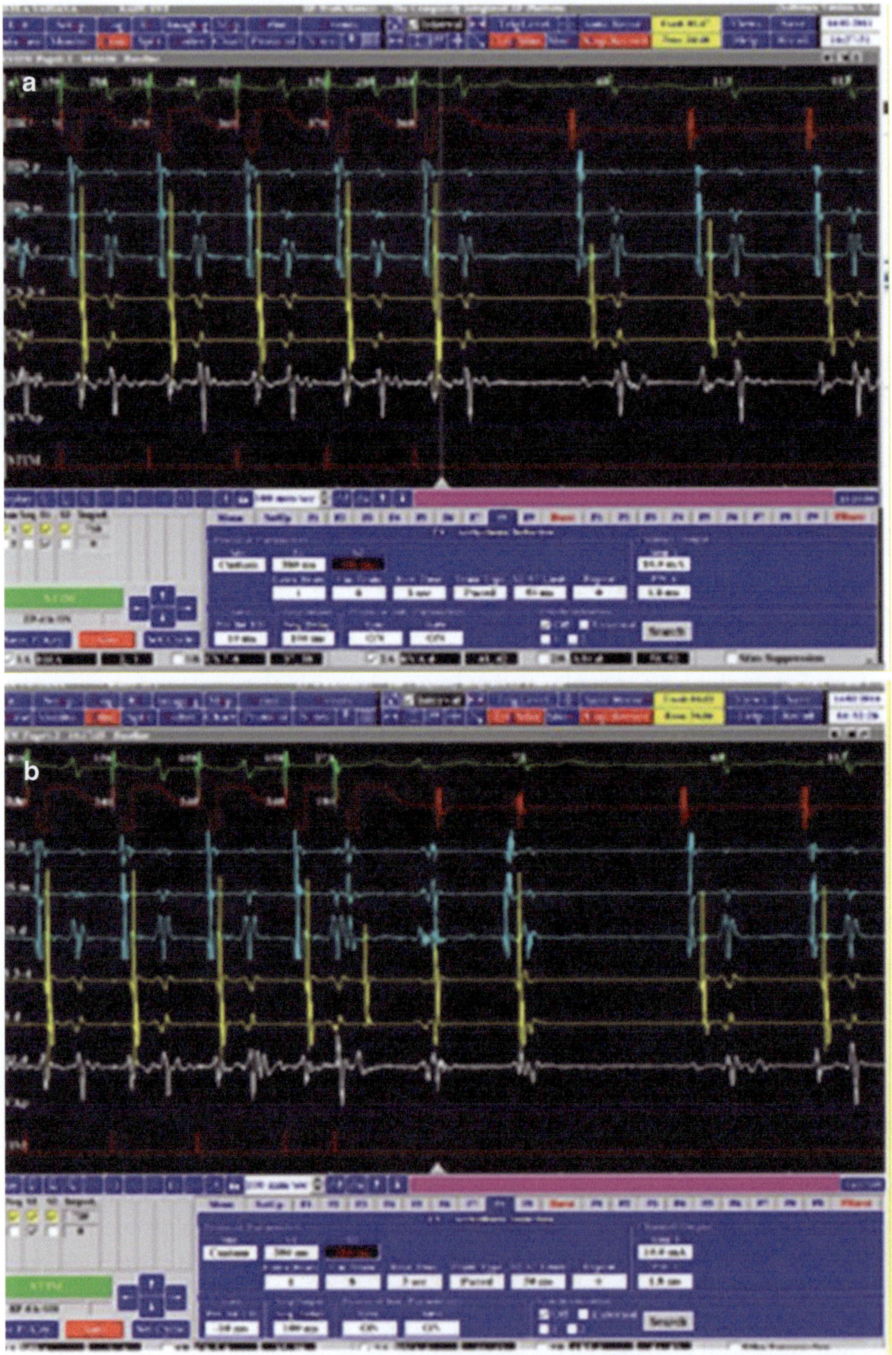

Fig. 11.1 Atrial stimulation for AVNRT induction without and with adrenaline. (**a**) Before adrenaline infusion, programmed atrial stimulation fails to induce atrial echo beats or AVNRT. (**b**) After adrenaline infusion, programmed atrial stimulation induces two atrial extra beats

Table 11.1 Comparison between the two groups: isoprenaline and atropine find no significant difference between them

Independent samples test										
		Levene's test for equality of variances		*t*-Test for equality of means						
									95% confidence interval of the difference	
		F	Sig.	*t*	df	Sig. (2-tailed)	Mean difference	Std. error difference	Lower	Upper
VAR00001	Equal variances assumed	3.341	0.072	−0.853	67	0.396	−0.08367	0.09804	−0.27937	0.11202
	Equal variances not assumed			−0.944	44.638	0.350	−0.08367	0.08866	−0.26228	0.09494

11.1.5 Discussion

11.1.5.1 Result Interpretation

Our study shows that both isoprenaline and adrenaline have a high sensibility for AVNRT induction: 81.6% for adrenaline and 90% for isoprenaline. The difference of 8.4% between the two drugs is nonsignificant, although it looks that isoprenaline would have a better profile. A higher number of patients are necessary to confirm the difference between the two drugs.

Adrenaline was generally well tolerated; the infusion had to be stopped in 4% of the patients (2 out of 49 patients). None of the patients from the isoprenaline group had significant side effects that would require termination of the infusion.

These results support the use of adrenaline in the EP lab for arrhythmia induction when isoprenaline is unavailable. The disadvantage of adrenaline is the increase in the blood pressure compared to isoprenaline which lowers the blood pressure; therefore some authors use the association of adrenaline with isoprenaline when the last one is unable to induce ventricular tachycardia [1].

We have found a similar efficacy for both isoprenaline and adrenaline. But the result does not mean that both drugs have the same efficacy as the low number of patients is a limitation of our study. For a significant p value and a difference of 5% between both groups, a number of 242 patients for adrenaline and isoprenaline would be mandatory.

11.1.5.2 Prior Studies with Isoprenaline

Hariman et al. [2] demonstrated that isoprenaline is effective for PSVT induction in patients that are non-inducible in the basal state or after atropine injection. Toda et al. [3] found a sensitivity of 50% for isoprenaline for the induction of paroxysmal supraventricular tachycardia in patients that had the arrhythmia during an effort. Huycke et al. [4] found a sensitivity of 67% and a specificity of 100% for PSVT induction in 20 patients with dual nodal pathway without any arrhythmia induced in the basal state. In the same year, Brembilla-Perot et al. [5] published their results with isoprenaline and showed similar results to our study, a sensitivity of 90% for induction of effort-induced PSVT (29 out of 32 patients). In the control group of 37 patients without PSVT, after isoprenaline infusion it was impossible to induce PSVT and allowed the calculation of a 100% specificity for isoprenaline. Stellbrink et al. [6] demonstrated that isoprenaline infusion in patients with AVNRT facilitated induction in 93% of the total of 80 patients. Katz et al. [7] reported a sensitivity of 46% for PSVT induction in patients with ventricular preexcitation. Oral et al. [8] demonstrated a sensitivity of 80% and a specificity of 95% for atrial fibrillation induction after 20 μg/min isoprenaline, in patients with paroxysmal persistent or long-term persistent atrial fibrillation.

11.1.5.3 Prior Studies with Adrenaline

The effect of adrenaline in patients with supraventricular tachycardias is known; it facilitates arrhythmia induction, but studies are made on a low number of patients and give no information on the sensibility of this drug for PSVT induction. Increase in plasma concentration of adrenaline leads to decrease in atrial refractory period, decrease in refractory period of the AV node, and increase in the conduction velocity through the AV node, facilitating induction of AVNRT or orthodromic reentrant tachycardia in patients with accessory pathways.

In the study of Cismaru et al. [9], an increase in the plasma concentration of adrenaline facilitated PSVT induction in patients that were non-inducible in the basal state.

11.1.5.4 Limitations

This study has two important limitations: first is a nonrandomized study that considered all the patients that received adrenaline and isoprenaline, even if we found significant differences between the two groups in terms of sex representation. Matching for age and sex between both groups could not be possible because a much lower number of patients would have been obtained. The number of patients in this study was already low and made the comparison between adrenaline and isoprenaline nonsignificant.

11.1.6 Conclusion

Adrenaline and isoprenaline showed a similar efficacy for AVNRT induction in the EP laboratory. Adrenaline can be used when isoprenaline is difficult to obtain. The calculated efficacy for adrenaline was 81.5 and 90% for isoprenaline, without significant statistical difference between them. To obtain a significant difference between the two drugs with a significant p value lower than 0.05, a higher number of patients is required in both groups.

11.2 Adrenaline Versus Isoprenaline for Ischemic VT Induction

11.2.1 Introduction

Ventricular tachycardia in patients with ischemic heart disease has a poor prognosis. In those patients the best option is an ICD implant for the prevention of sudden cardiac death. Catheter ablation is an auxiliary method that is performed in patients implanted with an ICD that received multiple internal electrical shocks. In some Eastern countries, ICDs are not available for all the patients that need it so catheter ablation, when available, can be performed in patients with episodes of VT without ICD protection.

In some patients clinical VT can be induced by stimulating the ventricle at the level of RVOT or RV apex. When VT is non-inducible, drugs are used in the EP lab to facilitate arrhythmia induction. The most used drug is isoprenaline which is considered the gold standard, but in some East European countries, the drug is not available. Therefore other options have been searched. Atropine is not a good option because its effect is mostly on the sinus node and AV node, with small or almost no effect on the atrial and ventricular myocardium. We tested adrenaline for VT induction, and we compared the results with those existing in the medical literature on isoprenaline.

11.2.2 Hypothesis

In order to compare isoprenaline with adrenaline, we took all the patients with ischemic VT that received adrenaline for VT induction and compared with reported sensibility of isoprenaline for VT induction found in medical literature. The effects of isoprenaline and adrenaline are slightly different as isoprenaline acts on β1 and β2 receptors and adrenaline acts also on α1 and α2 receptors.

11.2.3 Materials and Method

11.2.3.1 Study Group

Ninety-two consecutive patients with ventricular tachycardia were selected for this study, having (1) old myocardial infarction, (2) repeated episodes of ventricular tachycardia, (3) treated with amiodarone, and (4) performing an electrophysiological study. Of the 92 patients, we excluded those with RVOT, LVOT ventricular tachycardia, fascicular VT, patients with right ventricular dysplasia, left ventricular noncompaction, and left ventricular hypertrophy. In the end only 12 patients presented the inclusion criteria for further analysis.

11.2.3.2 Electrophysiological Study

All the patients came to the EP lab in the post-absorptive state. Amiodarone was continued the days before electrophysiological study and the same day because the effect of amiodarone was studied during intervention. The right femoral or right jugular vein was punctured, and one or two catheters were inserted at the level of right ventricular apex and RVOT. Ventricular stimulation was performed with up to three extra beats on an imposed rhythm of 600 and 400 ms. If no sustained arrhythmia was induced, lasting for more than 30 s or with hemodynamic instability, then adrenaline was infused.

11.2.3.3 Adrenaline Infusion

The dose of adrenaline infused was 0.05–0.2 μg/kg/min adjusted to obtain a heart rate over 100 bpm but not more than 150 bpm. During adrenaline infusion, programmed atrial and ventricular stimulation was performed at the apex of the right ventricle and RVOT.

There were no important ischemic side effects like angina or ST significant depression or elevation on 12-lead ECG. The most frequent side effect was tremor, which was well tolerated by the patients. The second encountered side effect was rise in the blood pressure, but after lowering the dose of adrenaline, the blood pressure normalized, but the heart rate remained over 100 bpm.

11.2.3.4 Catheter Mapping and Ablation

For catheter ablation the conventional approach was used with fluoroscopic guidance for the first cases and the three-dimensional mapping using the EnSite NAVX Saint Jude system or the CARTO-Biosense system.

When the conventional approach was used, a coronary sinus catheter was placed inside the coronary sinus to locate the mitral valve plane. The ablation catheter inside the left ventricle was an irrigated 4 mm tip one.

In every patient, attempts were made to induce ventricular tachycardia and to map the left ventricle during VT. If no VT could be induced during standard atrial and ventricular stimulation, then adrenaline was infused to facilitate induction. When VT was associated with hemodynamic instability, points of activation were taken during VT and points of voltage during sinus rhythm to make an idea about the reentry circuit.

Both antegrade and retrograde approaches were used to reach the left ventricle. Patients received i.v. heparin in dose of 70 U/kg and further doses to maintain an ACT of 250–350 ms. We did not use the classical approach with entrainment mapping because we believe it can change the cycle length of the tachycardia and change the activation pattern. So the value of the post-pacing interval was not useful in our patients. On the other hand, we used activation mapping to search for the reentry circuit and ablate at the level of the reentrant isthmus.

Catheter ablation was performed inside the left ventricle with 35–45 W and a cutoff temperature of 55 °C for irrigated catheters. Rarely non-irrigated catheters were used for the first cases of VT ablation.

11.2.3.5 Statistical Analysis

For statistical analysis the SPSS program was used. Normal distribution was tested using the Shapiro-Wilk test. In case of normal distributed values, the student test was used for comparison and in case on non-normal distributed values the Mann-Whitney U test.

11.2.4 Results

The studied population consisted in 12 patients with a mean age of 55 years old of which 85.7% were males. All patients had a history of myocardial infarction. In none of the 12 patients, ventricular tachycardia could be induced during programmed ventricular stimulation because patients were treated and protected by amiodarone (Figs. 11.2 and 11.3).

After adrenaline infusion in 6 of the 12 patients, ventricular tachycardia could be induced. The duration of the VT was either >30 s, or it associated hemodynamic instability so it was called sustained VT (Fig. 11.4). The mean cycle length of the VT was 315 ms with a minimum of 210 and a maximum of 400 ms. In six patients despite adrenaline, VT could not be induced as they were protected by amiodarone. Of the six patients, three had an old inferior myocardial infarction, and three had an old anterior myocardial infarction. The mean age in the group of patients with inducible VT was 56 years, and the mean age was 54 years in the group without induced VT under adrenaline (nonsignificant p value).

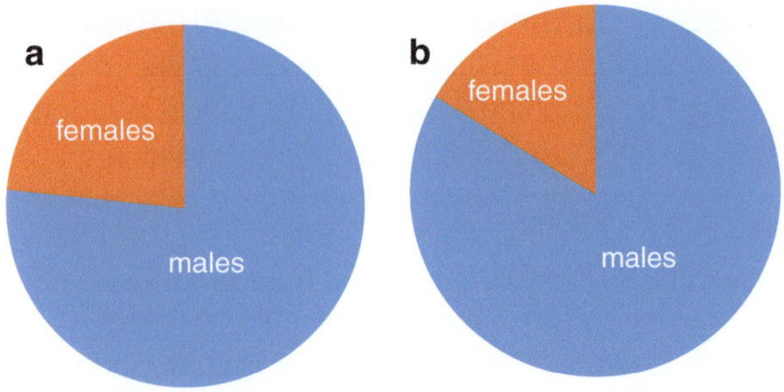

Fig. 11.2 Sex distribution in the (**a**) control group and (**b**) adrenaline group

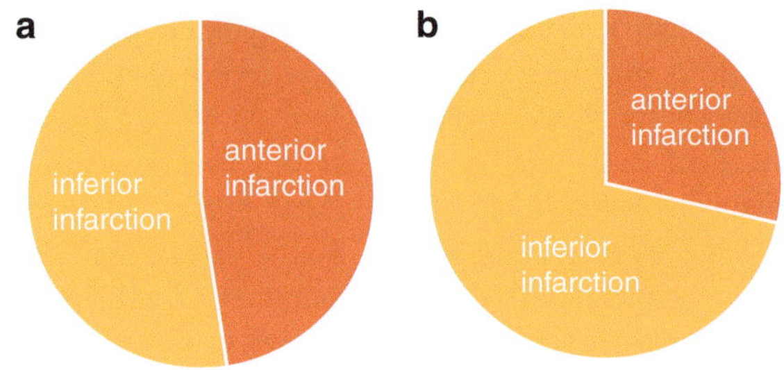

Fig. 11.3 Distribution of myocardial infarction in the two groups: (**a**) control group and (**b**) adrenaline group

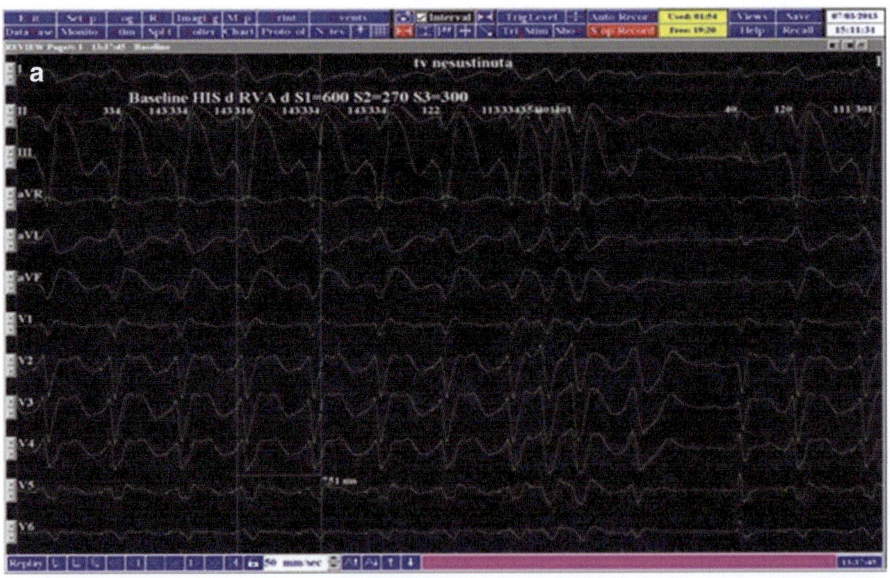

Fig. 11.4 Induction of ventricular tachycardia in a patient treated with amiodarone: (**a**) before adrenaline infusion the patient is protected by amiodarone, and no VT can be induced with programmed ventricular stimulation. (**b**) After adrenaline infusion in dose of 0.1 μg/kg/min, monomorphic VT is induced

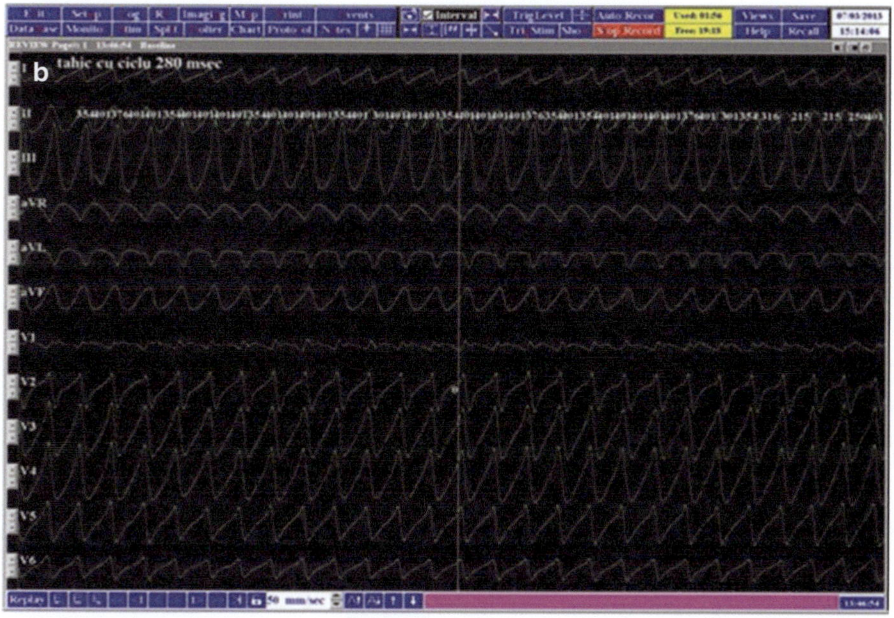

Fig. 11.4 (continued)

Of the five patients with inducible VT, mapping and ablation were performed using the CARTO 3 Biosense Webster system in two patients, the EnSite NAVX system in one patient and the conventional radioscopic system in one patient. For the last patient, as he presented monomorphic VT with hemodynamic instability, substrate ablation was performed with ICD implantation.

Adrenaline was infused in dose of 0.05–0.3 μg/kg/min with tremor as the only side effect during infusion. It was present in three patients but was well tolerated, and the infusion was continued despite this side effect.

No patient presented ST elevation on the 12-lead ECG. In the control group, there were 21 patients with the mean age of 60.7, males 76.2% that were treated with amiodarone and did not receive adrenaline for VT induction. Of the 21 patients, 11 had an old inferior myocardial infarction and 10 an old anterior myocardial infarction. The mean tachycardia cycle length was 275 ms.

Between the two groups, there were no differences in terms of localization of the myocardial infarction: anterior or inferior, ejection fraction, left ventricular dimensions, and tachycardia cycle length. The only difference was the inducibility of VT which was 50% in the adrenaline group and 100% in the amiodarone group (Fig. 11.5). The selection of patients was made to have all the patients inducible in the amiodarone group, without receiving adrenaline infusion.

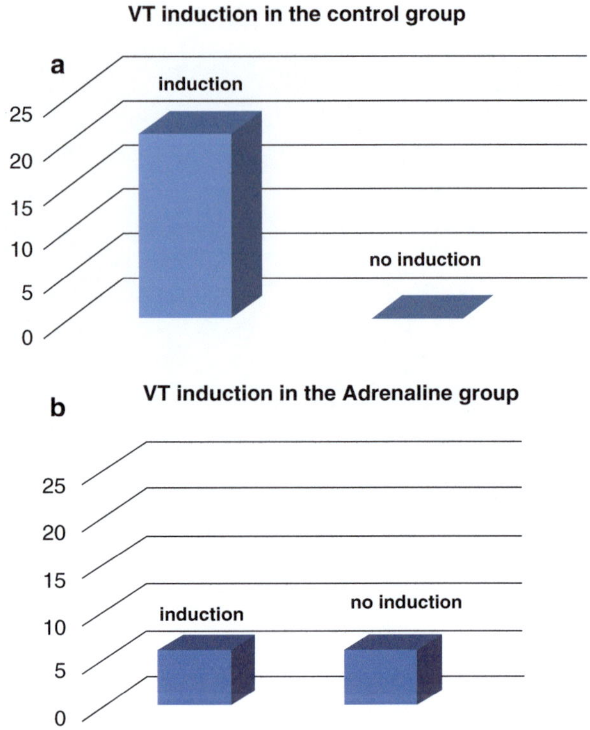

Fig. 11.5 VT induction in the control group and adrenaline group. (**a**) In the control group, VT was inducible in the basal state. (**b**) In the adrenaline group, VT was non-inducible in the basal state, and after adrenaline infusion, only 50% of the patients presented VT

11.2.5 Discussion

11.2.5.1 Main Findings

This study shows that adrenaline can be used in the EP lab for VT induction in patients that are already treated with amiodarone when VT is not inducible in basal conditions. Ischemic VT is the most frequent form of VT in patients with structural heart disease. Most of the time, these patients would be treated with amiodarone as they have a relative contraindication to class I antiarrhythmic drugs. In our patient population, the inducibility rate with adrenaline was 50%.

11.2.5.2 Other Studies with Adrenaline

Morady et al. verified inducibility of ventricular tachycardia in patients treated with adrenaline. They studied 21 patients with ischemic monomorphic ventricular tachycardia treated with quinidine. The infusion rate for adrenaline was 0.025 or 0.05 μg/kg/min. Of the 21 patients treated with quinidine, only 12 were "protected" by quinidine treatment and were uninducible during programed ventricular stimulation. After adrenaline infusion 2 of the 12 patients became inducible. Therefore the inducibility rate under adrenaline was 17%.

Calkins et al. compared the antiarrhythmic effects of quinidine and amiodarone in patients with ventricular tachycardia. The infusion rate for adrenaline was 0.025 μg/kg/min and 0.05 μg/kg/min. Five patients who received amiodarone were uninducible during programmed ventricular stimulation. After adrenaline infusion none of the five patients became inducible. The inducibility rate under adrenaline in this study was 0% [10].

11.2.5.3 Adrenaline Versus Isoprenaline

We compared studies with adrenaline and isoprenaline in terms of inducibility rate of VT.

Jasayeri et al. reported an induction rate of 59% in patients treated with class I antiarrhythmic drugs that received isoprenaline for VT induction. Patients in this study had coronary artery disease, dilated cardiomyopathy, or no structural heart disease [11].

In accordance, Markel et al. infused isoprenaline 0.03 μg/kg/min in five patients with ventricular tachycardia treated with procainamide, quinidine, propafenone, mexiletine, and lidocaine. Those antiarrhythmic drugs prolonged the cycle length of ventricular tachycardia, and isoprenaline infusion brought the cycle length to the initial value [12].

Compared to isoprenaline, adrenaline had a lower inducibility rate both in the studies of Morady and Calkins. In our study we found an inducibility rate of 50% which is also lower than the percentage found for isoprenaline. Therefore the best option for VT induction remains isoprenaline. Adrenaline should be reserved for cases where isoprenaline is unavailable. Another option like atropine is not a good alternative as the effect on the ventricles is very low with improbable induction of VT.

References

1. Choi EK, Kumar S, Nagashima K, Lin KY, Barbhaiya CR, Chinitz JS, et al. Better outcome of ablation for sustained outflow-tract ventricular tachycardia when tachycardia is inducible. Europace. 2015;17:1571–9.
2. Hariman RJ, Gomes JA, El-Sherif N. Catecholamine-dependent atrioventricular nodal reentrant tachycardia. Circulation. 1983;67:681–6.
3. Toda I, Kawahara T, Murakawa Y, Nozaki A, Kawakubo K, Inoue H, et al. Electrophysiologial study of young patients with exercise related paroxysmal of palpitation: role of atropine and isoprenaline for initiation of supraventricular tachycardia. Br Heart J. 1989;61:268–73.
4. Huycke EC, Lai WT, Nguyen NX, Keung EC, Sung RJ. Role of intravenous isoproterenol in the electrophysiological induction of atrioventricular node reentrant tachycardia in patients with dual atrioventricular node pathways. Am J Cardiol. 1989;64:1131–7.
5. Brembilla-Perrot B, Terrier de la Chaise A, Pichene M, Aliot E, Cherrier F, Pernot C. Isoprenaline as an aid to the induction of catecholamine dependent supraventricular tachycardia during programmed stimulation. Br Heart J. 1989;61:348–55.
6. Stellbrink C, Diem B, Schauerte P, Brehmer K, Schuett H, Hanrath P. Differential effects of atropine and isoproterenol on inducibility of atrioventricular nodal reentrant tachycardia. J Interv Card Electrophysiol. 2001;5:463–9.
7. Katz DF, Hegeman TW, Aleong R. Use of isoproterenol for assessment of anterograde accessory pathway conduction properties in sedated asymptomatic patients with ventricular pre-excitation. J Innov Card Rhythm Manage. 2013;4:1428–33.
8. Oral H, Crawford T, Frederick M, Gadeela N, Wimmer A, Dey S, et al. Inducibility of paroxysmal atrial fibrillation by isoproterenol and its relation to the mode of onset of atrial fibrillation. J Cardiovasc Electrophysiol. 2008;19:466–70.

9. Cismaru G, Rosu R, Muresan L, Puiu M, Andronache M, Hengan E, et al. The value of adrenaline in the induction of supraventricular tachycardia in the EP lab. Europace. 2014;16:1634–8.
10. Calkins H, Sousa J, El-Atassi R, Schamltz S, Kadish A, Morady F. Reversal of antiarrhythmic drug effects by epinephrine: quinidine versus amiodarone. Am J Cardiol. 1992;19:347–52.
11. Jazayeri M, VanWyhe G, Avitall B, McKinnie J, Tchou P, Akhtar M. Isoproterenol reversal of antiarrhythmic effects in patients with inducible sustained ventricular tachyarrhythmias. J Am Coll Cardiol. 1989;14:705–11.
12. Markel ML, Miles WM, Prytowsky EN. The differential effects of isoproterenol on sustained ventricular tachycardia before and during antiarrhythmic therapy. J Am Coll Cardiol. 1987;9:227–8.

How to Induce Arrhythmias with Salbutamol

12

Piotr Futyma

12.1 Introduction

Salbutamol is a selective beta2 adrenoreceptor agonist causing bronchodilation. It is commonly used for the treatment of acute bronchoconstriction episodes such as acute asthmatic crisis or status asthmaticus. It is also used for treatment of chronic obstructive pulmonary disease. Salbutamol is the sulphate salt of the albuterol, a phenethylamine with bronchodilator properties (Fig. 12.1). As a β2 selective agonist, intravenous salbutamol may also serve as a pharmacologic agent for the induction of arrhythmias during electrophysiological study (EPS). At St. Joseph's Heart Center, Rzeszów, salbutamol (Salbutamol WZF, Polfa Warszawa S.A.) is a main drug used during EPS in case of arrhythmia non-inducibility.

P. Futyma (✉)
Invasive Cardiology Department, St. Joseph's Heart Center, Rzeszow, Poland

G. Cismaru (ed.), *Arrhythmia Induction in the EP Lab*,
https://doi.org/10.1007/978-3-319-92729-9_12

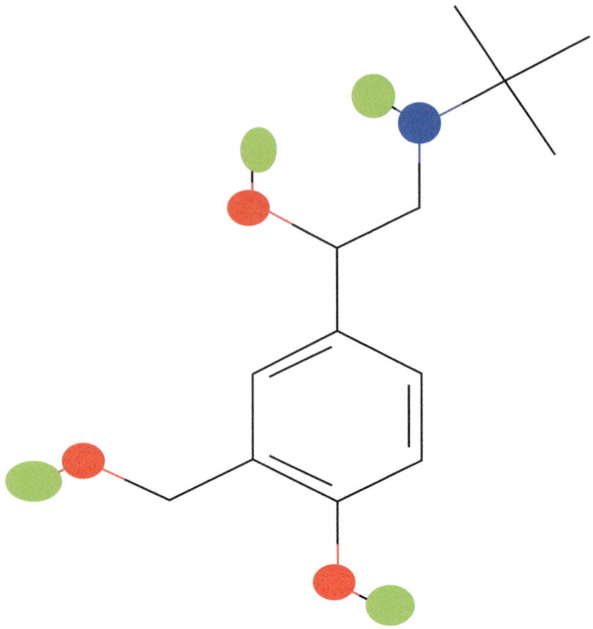

Fig. 12.1 Chemical structure of salbutamol: C13, H21, NO3

12.2 Electrophysiological Effects

Infusion of salbutamol significantly shortens sinus node cycle length and sinus node recovery time. It also shortens effective refractory period of the AV node, significantly shortens AH interval and the Wenckebach point (WP), and decreases atrial and ventricular effective refractory periods [1–5]. Salbutamol is more likely to have impact on sinus node's activity compared to the AV node—it was shown that the presence of β2 adrenoreceptors is more common in the sinoatrial node than in the right atrium [6–8].

12.3 Doses of Salbutamol

At low-infusion rates, salbutamol produces peripheral vasodilatation with a decrease in diastolic blood pressure and reflex increase in norepinephrine plasma concentration and sympathetic cardiac activity.

12.3.1 In Acute Asthmatic Crisis

After *salbutamol's nebulization*, the onset of action is present within 15 min. *Salbutamol's infusion* recommended doses:

- An initial bolus of 15 μg/kg (maximum 250 μg) over 15 min

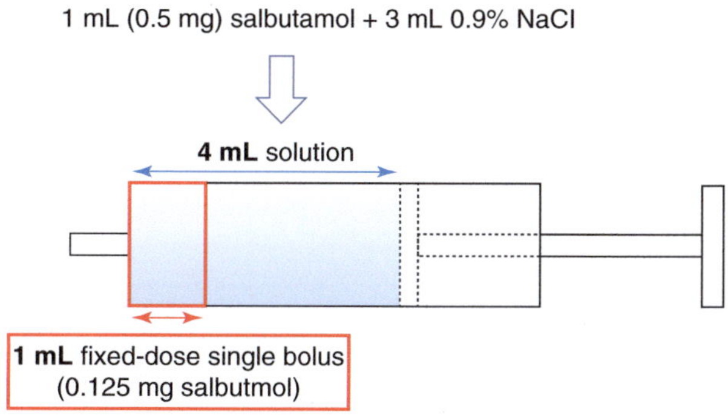

Fig. 12.2 Syringe preparation

- An Intravenous infusion dose of 1–2 µg/kg/min. Intravenous salbutamol can be diluted with 5% glucose or 0.9% sodium chloride.

12.3.2 In the EP Lab

Doses used in the setting of EPS differ from those used during acute asthmatic crisis. A recommended single bolus dose for use in the EP lab equals 0.125 mg of salbutamol.

12.3.2.1 Infusion Preparation

1. The patient preparation should be conducted following standard EP lab procedure regimens. Patient should undergo fasting and IV access should be established prior to induction.
2. An infusion pump is NOT necessary for salbutamol administration.
3. Mix 1 mL (0.5 mg) of salbutamol with 3 mL of 0.9% NaCl into the syringe (Fig. 12.2).
4. A single, fixed dose consists of 1 mL solution (0.125 mg of salbutamol).
5. After a single bolus is administered, the heart rate should increase, usually within 20–60 s programmed stimulation can be performed when heart rate reaches >130% of baseline (Fig. 12.3).
6. Intravenous administration of salbutamol can be repeated when necessary (i.e., in case of limited effect of a single-dose bolus).
7. After ablation, administration of salbutamol can be repeated when appropriate.
8. After the EP study, potassium levels should be checked (as salbutamol induces intracellular increase of potassium with subsequent decrease of potassium levels in blood serum).
9. Half-life elimination of salbutamol ranges from 4 to 6 h.

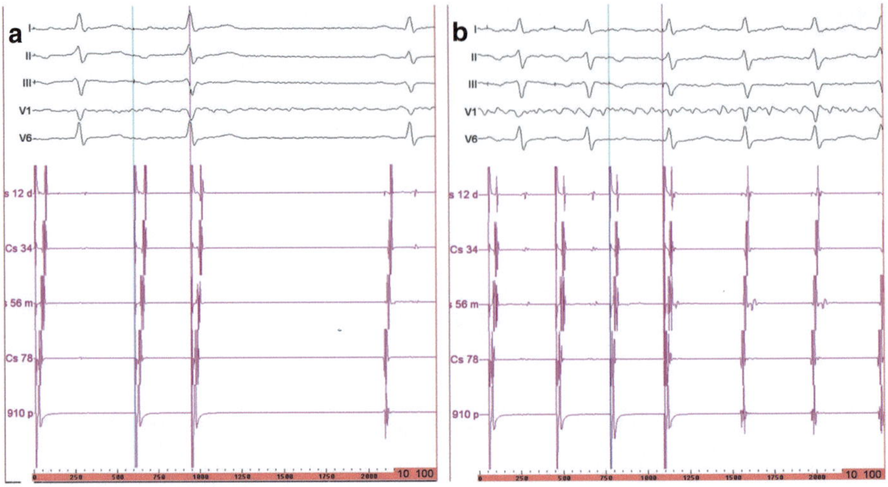

Fig. 12.3 Non-inducibility at baseline (**a**) and induction of typical AVNRT after single fixed dose of intravenous salbutamol (**b**)

12.4 Contraindications and Precautions for Salbutamol Use During EPS

Contraindications

- Intolerance of salbutamol or any of the intravenous ingredients
- Age <6 years

Precautions

- Heart failure
- Ischemic heart disease
- Hemodynamically significant left ventricular outflow tract obstruction
- Hypertension
- Hypokalemia
- Aneurysm
- Hyperthyroidism
- Pheochromocytoma
- Diabetes
- Intolerance of sympathomimetic drugs

12.5 Salbutamol's Possible Side Effects

Most common, minor side effects:

- Tremor
- Anxiety

- Headache
- Muscle cramps, muscle tremors (especially hands)
- Dry mouth
- Palpitation
- Nausea, vomiting
- Agitation, hyperactivity
- Feeling of warmth during bolus administration
- Hypotension caused by peripheral vasodilation
- Hyperglycemia
- Hypophosphatemia

Rare, major side effects

- Angioedema
- Bronchial spasm
- Hypotension
- Shock

12.6 Clinical Use

Literature on the use of salbutamol during EPS is limited. However, in cases of arrhythmia non-inducibility, and when isoprenaline is unavailable, salbutamol can be easily adapted for EP testing. It may also increase the sensitivity of EPS directly after catheter ablation. A few examples of relations between salbutamol use and arrhythmia occurrence were described in the literature.

Al-Hillawi et al. demonstrated that some patients who were treated with salbutamol or serbutaline could later develop arrhythmias such as monomorphic or bifocal ventricular premature contractions and short runs of paroxysmal atrial tachycardia [9]. Trachsel et al. reported two cases of salbutamol-induced supraventricular tachycardia (SVT) successfully converted to sinus rhythm after adenosine administration 0.1 mg/kg, without worsening of asthma symptoms [10].

A case of SVT after standard administration of salbutamol was also reported [11]—an 8-year-old patient receiving salbutamol developed an episode of orthodromic reentrant tachycardia (ORT) of 240 bpm. Two other cases in the pediatric population were also reported by Keller [12] and Cook [13]. The first case was a 19-month-old infant with salbutamol-induced SVT, converted to sinus rhythm after exposure of the infant's face to iced water. The second case was a 4-year-old child with salbutamol-induced SVT, converted to sinus rhythm after intravenous adenosine. Although adenosine has the potential of bronchospasm induction in asthmatic patients and was safely administrated in the presented case, its use requires cautiousness in former patients.

Refractory SVT can also occur as a result of salbutamol toxicity, as previously reported [14].

In patients with Wolf-Parkinson-White syndrome a nebulized salbutamol can induce ORT several minutes after administration [15, 16]. Attempts however to induce ORT with salbutamol infusion only, without the need of programmed stimulation, may fail [15].

Also other supraventricular arrhythmias can be enhanced by salbutamol. Induction of focal atrial tachycardia was also documented previously [17].

High doses of salbutamol can lead to ventricular fibrillation. Uysal et al. reported a case of 24-year-old female, who developed hypokalemia 2.3 mmol/L and subsequent ventricular fibrillation after suicide attempt with high dose of salbutamol (76 mg) [18].

12.7 How to Induce Arrhythmias in the EP Lab

At St. Joseph's Heart Center, the main indication for administration of salbutamol during EPS is when SVT cannot be induced with programmed stimulation at baseline. When SVT suspicion is high, a single bolus of salbutamol can lead to induction of clinical SVT and subsequent appropriate treatment of arrhythmia. Due to sinus tachycardia, which usually follows a single dose of salbutamol, and due to improvement of conduction properties of atrioventricular node, more aggressive stimulation protocols should be delivered, usually with up to triple extrastimuli on an eight-beat drive train (Fig. 12.3).

Salbutamol can also be used for induction of premature ventricular complexes (PVCs). Activation mapping has a higher value for PVC localization than pacemapping, and this can be facilitated by salbutamol infusion. Induction of PVCs may occur directly after salbutamol infusion, sometimes boosted with programmed ventricular or atrial stimulation, or within minutes after the increase of heart rate was obtained ("the washout" phenomenon) (Fig. 12.4).

Concealed or intermittent preexcitation patterns may also be revealed in patients without visible preexcitation on ECG at rest. After ablation of an accessory pathway, retrograde conduction can also be verified after infusion of salbutamol.

Adverse events of salbutamol administration at the time of EPS may occur very rarely, but have mostly been documented to be palpitations. Sinus tachycardia may last within hours after EPS and can be well controlled with small doses of beta-blockers.

The use of salbutamol for EP testing can be an attractive alternative when other drugs are unavailable. Its electrophysiological utility should be confirmed in future studies.

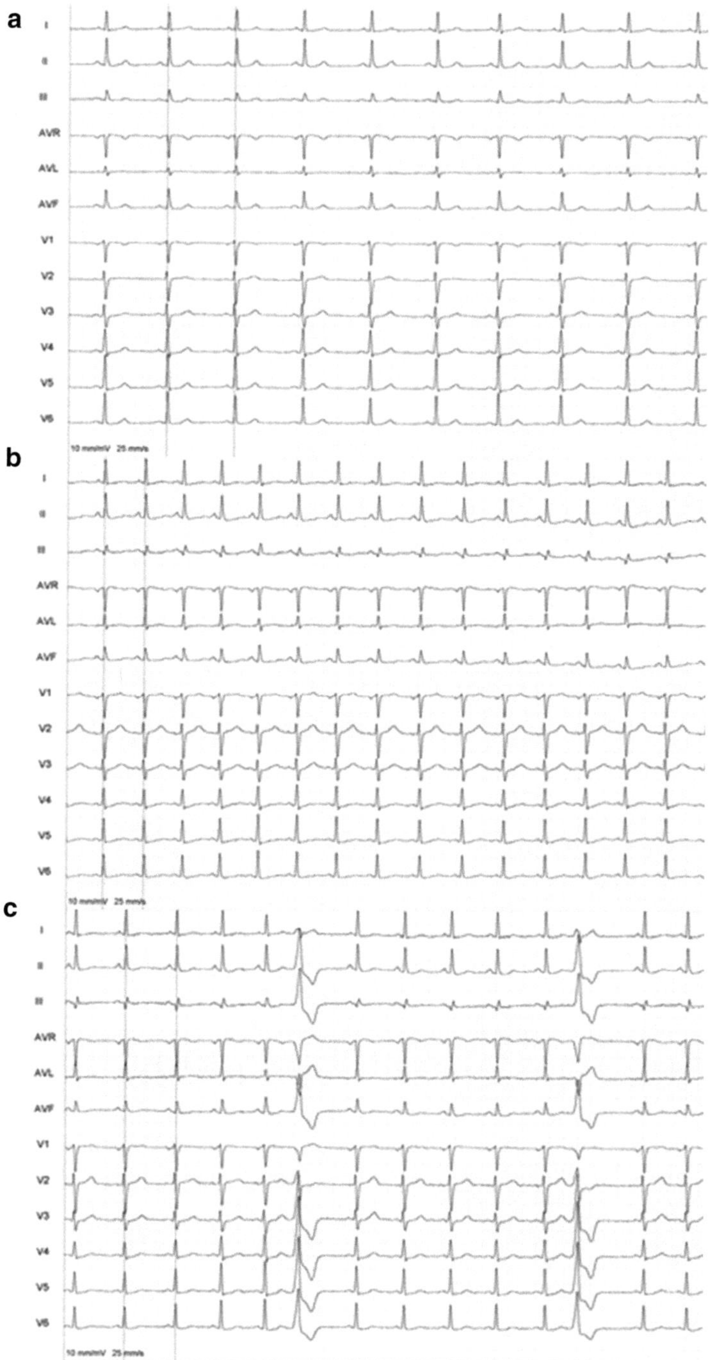

Fig. 12.4 Example of a "washout" phenomenon. Panel **a**: baseline ECG with no PVCs, Panel **b**: sinus tachycardia after single dose of salbutamol, Panel **c**: after heart rate decreases, clinical PVCs show up

References

1. Davis E, Loiacono R, Summers RJ. The rush to adrenaline: drugs in sport acting on the β-adrenergic system. Br J Pharmacol. 2008;154(3):584–97.
2. Cekici L, Valipour A, Kohansal R, Burghuber OC. Short-term effects of inhaled salbutamol on autonomic cardiovascular control in healthy subjects: a placebo-controlled study. Br J Clin Pharmacol. 2009;67(4):394–402.
3. Selective beta2 agonists—side effects. British National Formulary. 57 ed. London: BMJ Publishing Group Ltd and Royal Pharmaceutical Society Publishing; 2008. ISBN 0-85369-778-7.
4. Lowe MD, Rowland E, Brown MJ, Grace AA. Beta(2) adrenergic receptors mediate important electrophysiological effects in human ventricular myocardium. Heart. 2001;86(1):45–51.
5. Insulander P, Juhlin-Dannfelt A, Freyschuss U, Valln H. Electrophysiologic effects of Salbutamol a beta 2 selective agonist. J Cardiovasc Electrophysiol. 2004;15:316–22.
6. Rodefeld MD, Beau SL, Schuessler RB, et al. Betaadrenergic and muscarinic cholinergic receptor densities in the human sinoatrial node identification of a high beta 2 adrenergic receptor density. J Cardiovasc Electrophysiol. 1996;7:1039–49.
7. Summers RJ, Molnaar P, Russel F, Elnatan J, Jones CR, Buxton BF, Chang V, Hmbley J. Coexistence and localization of beta 1 and beta 2 adrenoreceptors in the human heart. Eur Heart J. 1989;10:11–21.
8. Kallergs EM, Manios EG, Kanoupakis EM, Schiza SE, Mavrakis HE, Klapsinos NK, et al. Acute electrophysiologic effects of inhaled salbutamol in humans. Chest. 2005;127:2057–63.
9. Al-Hillawi AH. Incidence of cardiac arrhythmias in patients taking slow release salbutamol and terbutaline for asthma. Br Med J. 1984;287:863–7.
10. Trachsel D, Newth CJL, Hammer J. Adenosine for salbutamol-induced supraventricular tachycardia. Intensive Care Med. 2007;22:1676.
11. Duane M, Chandran L, Morelli PJ. Recurrent supraventricular tachycardia as a complication of nebulized albuterol treatment. Clin Pediatr (Phila). 2000;39(11):673–7.
12. Keller KA, Bhisitkul DM. Supraventricular tachycardia: a complication of nebulized albuterol. Pediatr Emerg Care. 1995;11:98–9.
13. Cook P, Scarfone RJ, Cook RT. Adenosine in the termination of albuterol-induced supraventricular tachycardia. Ann Emerg Med. 1994;24:316–9.
14. Say B, Degirmencioglu H, Kutman HGK, Uras N, Dilmen U. Taquicardia supraventricular en un recien nacido despues del tratamiento con salbutamol nebulizado. A proposito de un caso. Arch Argent Pediatr. 2015;113:e98–100.
15. Kroesen M, Maseland M, Smal J, Reimer A, van Setten P. Probable association of achyarrhythmia with nebulized albuterol in a child with previously subclinical wolff Parkinson White syndrome. J Pediatr Pharmacol Ther. 2012;17:93–7.
16. Bonnin AJ, Richmond GW, Musto PK, Volgman AS, Moy JN. Repeated inhalation of nebulized albuterol did not induce arrhythmias in a patient with Wolff-Parkinosn-White syndrome and asthma. Chest. 1993;103:1892–4.
17. Tandeter H, Kobal S, Katz A. Swallowing-induced atrial tachyarrhythmia triggered by salbutamol: case report and review of the literature. ClinCardiol. 2010;33(6):E116–20.
18. Uysal E, Solak S, Carus M, Uzun N, Cevik E. Salbutamol abuse is associated with ventricular fibrillation. Turk J Emerg Med. 2015;15:87–9.

How to Induce Arrhythmias with Caffeine?

13

Cecilia Lazea

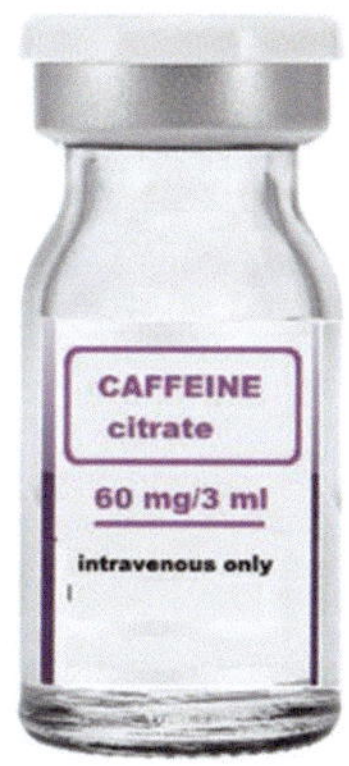

13.1 Introduction

Caffeine is the most used stimulant for the central nervous system. It is a natural methylxanthine that is present in many beverages (Fig. 13.1). In the USA 80% of adults use 3.5 cups of coffee/day on average the equivalent of 164 mg caffeine [1].

Caffeine can be found in coffee beans, tea leaves, kola nuts, and coca pods. Coffee beans also contain traces of theophylline. Coca-Cola was originally made from kola nut extract which is a source of caffeine although in early stages Coca-Cola also contained cocaine. Coke, Pepsi, Mountain Dew, and Dr. Pepper also contain 35–55 mg of caffeine per can, and the quantity is even higher for Red Bull and Rockstar with 80 mg per serving (Fig. 13.2).

C. Lazea (✉)
"Iuliu Hatieganu University of Medicine and Pharmacy", Pediatric Clinic No 1, Pediatric Cardiology, Cluj-Napoca, Romania
e-mail: cecilialazea@umfcluj.ro

G. Cismaru (ed.), *Arrhythmia Induction in the EP Lab*,
https://doi.org/10.1007/978-3-319-92729-9_13

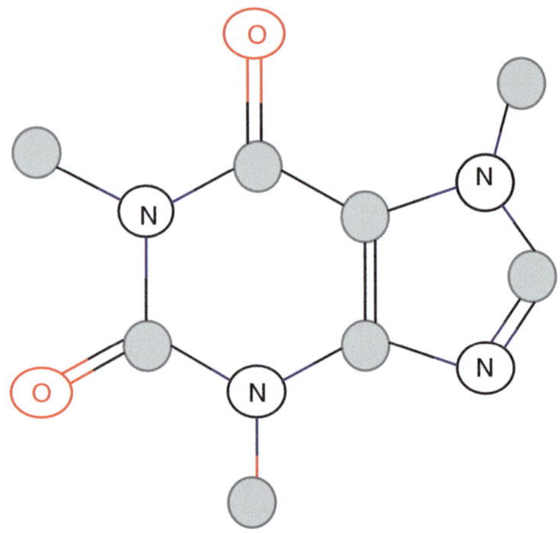

Fig. 13.1 Chemical structure of caffeine

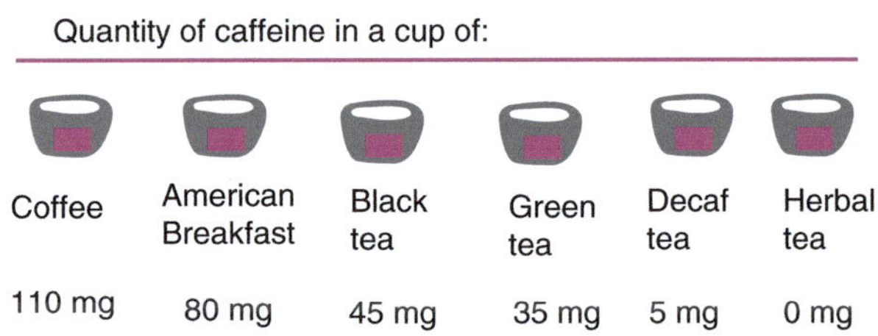

Fig. 13.2 Caffeine quantity in a cup of tea or coffee

13.2 Pharmacology

Biological half-life of caffeine is 3–7 h in adults, with a plasma half-life of 3–5 h. The pharmacological effects of caffeine are similar to other methylxanthines that are found in chocolate or tea. The fatal oral dose of caffeine is 10–14 g (150–200 mg/kg) or 75–100 regular cups of coffee [2]. Caffeine is rapidly absorbed within 45 min after ingestion [3, 4] with peak plasma concentrations occurring between 15 and 200 min after ingestion. The main pathway for caffeine metabolism is conversion to paraxanthine [5] which has no toxic effects in small doses. Both caffeine and paraxanthine increase the plasma concentration of epinephrine [6].

13.3 Mechanism of Action

Caffeine increases the action of norepinephrine in the human body and also augments the inotropic effects of other β-adrenergic drugs. It increases the heart rate, left ventricular output, and stroke output. It was demonstrated that a dose of 100 mg can increase alertness while a dose of 250 mg caffeine can increase the blood pressure [7].

Caffeine's action is mediated by several mechanisms: antagonization of adenosine receptors, inhibition of phosphodiesterase, release of calcium from intracellular stores, and antagonization of benzodiazepine receptors [8].

Due to its action on adenosine receptors, caffeine affects the release of norepinephrine, dopamine, and acetylcholine, with augmented effects on cardiac arrhythmogenesis [9].

Caffeine increases intracellular concentrations (cAMP) by phosphodiesterase inhibition. Increased cAMP leads to an increase in plasma catecholamines with cardiostimulatory and antiasthmatic actions.

Caffeine in high concentrations increases myofilamental sensitivity to calcium [10] and induces storage of calcium in the sarcoplasmic reticulum [11].

The action of caffeine on the sodium-potassium-ATP pump leads to a decrease in plasma concentration of potassium [12].

Robertson et al. [13] demonstrated a significant increase in plasma concentrations of catecholamines after caffeine ingestion. These results were confirmed by the work of Izzo et al. [14].

13.4 Doses of Caffeine

Caffeine can be injected intravenously in dose of 500 mg (250 mg caffeine and 250 mg sodium benzoate/2 mL). Data regarding caffeine use come from studies on patients with postdural lumbar puncture headache. Daily dose in this case is 500 mg two times/day.

When caffeine benzoate or caffeine citrate is unavailable, caffeine shots or cups of brewed coffee can be used instead.

In our cardiology department we mostly use caffeine for ventricular premature contractions ablation. We induce ventricular premature beats by asking the patient to drink four cups of brewed coffee (4 × 120 mg), 1 h before the electrophysiological study. Approximately 30 min after ingestion, the number of ventricular premature beats increases, and we are able to perform activation mapping (Fig. 13.3, Tables 13.1 and 13.2).

For other types of arrhythmias, we could not find a clear advantage of caffeine over atropine, adrenaline, or isoprenaline.

13.5 Preparation

Fig. 13.3 One hour before the electrophysiological study, patients are invited to drink four cups of brewed coffee

Table 13.1 Infusion preparation for dopamine

1. The patient should not eat for at least 8 h before the electrophysiological study
2. Patients are instructed to stop medication for five times the halftime prior to electrophysiological testing unless instructed otherwise
3. One hour before the electrophysiological study, the patient drinks four cups of brewed coffee containing at least 120 mg caffeine each
4. If premature ventricular contractions are not evident after caffeine, then 240 mg of aminophylline is injected intravenously
5. In case of no effect after caffeine and aminophylline, then adrenaline is infused

Table 13.2 Protocol for caffeine administration before the electrophysiological study

Time	Dose
Start	1 cup of brewed coffee
5 min	1 cup of brewed coffee
10 min	1 cup of brewed coffee
15 min	1 cup of brewed coffee
After 1 h	EP study takes place

Table 13.3 Contraindications for caffeine

• Allergy to caffeine
• Renal impairment
• Hepatic impairment

13.6 Side Effects of Caffeine

Caffeine overdose results in altered mental state, nausea, restlessness, vomiting, anxiety, dizziness, diaphoresis, arrhythmias, opisthotonus, myoclonic jerks, muscle twitching, seizures, and even myocardial infarction [15].

Most frequent side effects that appear at our usual dose of approximately 500 mg are agitation and tremor. These side effects permit the continuation of the EP study because they are well tolerated by the patient (Table 13.3, Fig. 13.4).

13.7 Electrophysiological Effects of Caffeine

13.7.1 Animal Studies

Bellet et al. demonstrated in the early 1970s a decrease in the ventricular fibrillation threshold after caffeine ingestion. In his experiment, a dose of 12.5 mg/kg caffeine was injected in dogs, and the VF threshold was verified before and after injection. Caffeine decreased the VF threshold both in nonischemic and ischemic dogs [16].

Ishida et al. demonstrated that high-dose infusion of caffeine (1 mg/kg/min) is associated with ventricular ectopic beats in rabbits [17].

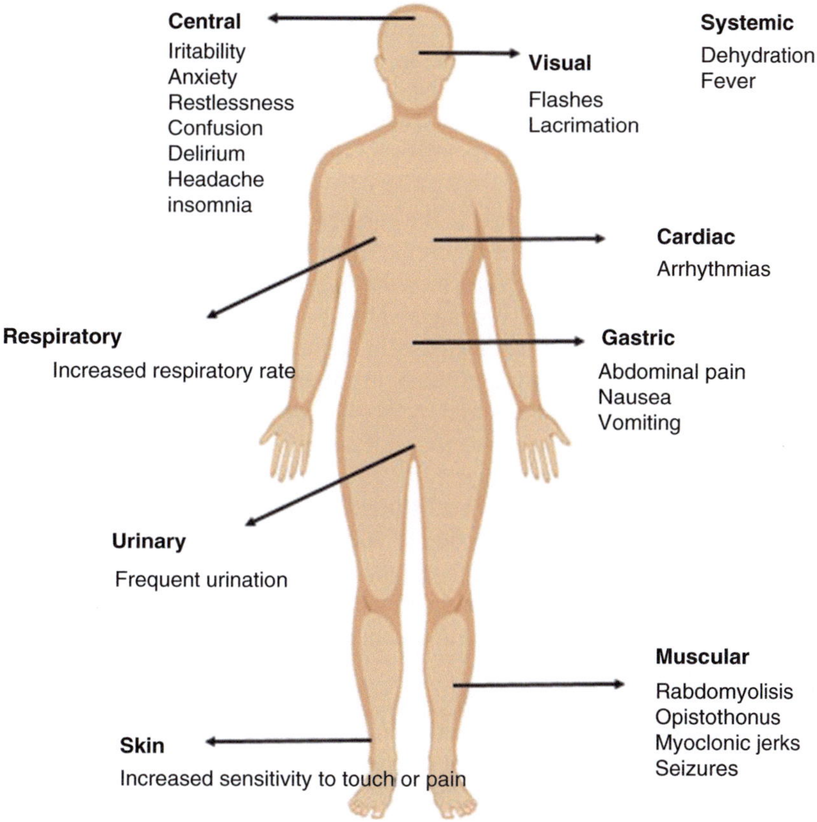

Fig. 13.4 Side effects of caffeine

Mehta et al. have shown a dose-response effect of caffeine on the occurrence of VPBs in dogs [18].

Many animal models showed a significant association between caffeine consumption and vulnerability to atrial fibrillation [16].

Rashid et al. showed that the period leading to the arrhythmia decreased with increasing plasma concentration of caffeine [19].

13.7.2 Human Studies

The effects of caffeine range from a negative chronotropic and inotropic effect in case of a low dose to a positive chronotropic and inotropic effect in case of a high dose and atrial and ventricular arrhythmias in case of very high dose [20].

Bioh et al. reported a case of 50 g caffeine intoxication in a 27-year-old woman with a history of depression. She presented a large QRS tachycardia, followed by multiple episodes of ventricular fibrillation. After continuous veno-venous hemofiltration, arrhythmic episodes stopped [21].

In the study of Chelsky et al., 275 mg of caffeine resulted in a plasma concentration of 6.2 μg/mL. The authors found no effect on right ventricular refractoriness after caffeine administration [22].

Prineas et al. showed an increase in the number of ventricular premature beats in subjects that consumed up to nine cups of coffee [12].

Sutherland et al. demonstrated an increase in the number of ventricular premature beats after 1 mg/kg coffee, as evaluated by 24-h Holter EKG [23].

Harris et al. also showed an increase in number and severity of ventricular premature beats after 900 mg/day caffeine, in patients with structural heart disease and low ejection fraction, as measured by 24–48-h Holter EKG [24].

Another study that evaluated the effects of caffeine using the electrophysiological study was that of Dobmeyer et al. [25]. They showed that RV refractory period decreased after caffeine ingestion in 12 patients. During baseline EP study, no VT was induced, but after caffeine ingestion of 200 mg, two patients presented nonsustained VT at programmed ventricular stimulation [26]. There was also a small but significant decrease in the refractoriness of the right atrium, right ventricle, and AV node of 10–20 ms. The mean serum concentrations after oral caffeine were 3.2 ± 0.4 μg/mL.

13.7.2.1 Caffeine for SVT Induction

Lemery et al. performed a placebo-controlled prospective randomized study in patients with supraventricular tachycardia undergoing an electrophysiological study. They found no association between a caffeine intake of 5 mg/kg and heart rate, atrial and ventricular refractoriness, conduction time through the AV node, or tachycardia inducibility [27]. This is in contrast with the study of Gould et al. who demonstrated significant decrease in the refractory period of the AV node after 150 mg of caffeine [28].

13.7.2.2 Caffeine for AF Induction

Rashid A et al. demonstrated on seven dogs that 1 mg/kg caffeine administered intravenously reduced the propensity for atrial fibrillation [19].

Di Rocco et al. showed that excessive caffeine consumption induced atrial fibrillation in two teenagers after energy drink consumption [29].

Frost and Vestergaard demonstrated in a large observational study of 50,000 individuals that caffeine is not associated with an increased incidence of atrial fibrillation. Only 555 individuals developed atrial fibrillation or flutter during 6 years follow-up, a value without statistical significance [30].

13.7.2.3 Caffeine for VT Induction

Chelsky et al. evaluated the effect of caffeine on programmed ventricular stimulation in 22 patients with a history of ventricular tachycardia or ventricular fibrillation. In order to verify the effects of caffeine, patients underwent an EP study before and 1 h after coffee consumption (500 mL coffee over 15-min period = 275 mg caffeine). In a regular instant coffee we can find around 60 mg caffeine after stirring two teaspoons of instant coffee in a cup of hot water. The plasma caffeine level was 6.2 ± 0.5 mg/L. Ten patients (46%) showed no change in VT inducibility after caffeine ingestion. Six patients (27%) had easier induction of VT after caffeine ingestion (decrease in the number of extrastimuli required to induce VT). Arrhythmia induction had no relationship with catecholamine or potassium plasma concentrations [22]. However, Graboys et al. showed an increase in serum caffeine levels after 200 mg caffeine drink, without any increase in VT, VF or non-sustained VT episodes [24].

References

1. Raebel M, Black J. The caffeine controversy: what are the facts? Hosp Pharm. 1984;19:257–67.
2. Hodgamn MJ. Caffeine. In: Wexler P, editor. Encyclopedia of toxicology. San Diego: Academic Press; 1988. p. 209–10.
3. Bonati M, Latini R, Galleti F. Caffeine disposition after oral doses. Clin Pharmacol Therap. 1982;32:98–106.
4. Liguori A, Hughes JR, Grass JA. Absorption and subjective effects of caffeine from coffee, cola and capsules. Pharmacol Biochem Behav. 1997;58:721–6.
5. Stavric B. Methylxanthines: toxicity to humans. 2. Caffeine. Food Chem Toxicol. 1988;26:645–62.
6. Watson J, Deary I, Kerr D. Central and peripheral effects of sustained caffeine use: tolerance is incomplete. Br J Clin Pharmacol. 2002;54:400–6.
7. Nawrot P, Jordan S, Eastwood J, Rotstein J, Hugenholtz A, Feeley M. Effects of caffeine on human health. Food Addit Contam. 2003;20:1–30.
8. Myers JP, Johnson DA, McVey DE. Caffeine and the modulation of brain function. In: Gupta BS, Gupta U, editors. Caffeine and behaviour: current views and research trends. Boca Raton: CRC Press; 1999. p. 17–30.
9. Daly JW, Shi D, Nikodyivic O, Jacobson KA. The role of adenosine receptors in the central action of caffeine. In: Gupta BS, Gupta U, editors. Caffeine and behaviour: current views and research trends. Boca Raton: CRC Press; 1999. p. 1–16.
10. Daly JW. In: Garattini S, editor. Caffeine, coffee, and health Mechanism of action of caffeine. New York: Raven Press; 1993. p. 97–150.
11. Nehlig A, Daval JL, Derby G. Caffeine and the central nervous system: mechanisms and action, biochemical, and psychostimulant effects. Brain Res Rev. 1992;17:139–70.
12. Prineas R, Jacobs D, Crow R, Blackburn H. Coffee, tea and VPB. J Chronic Dis. 1980;33:67–72.
13. Robertson D, Wade D, Workman R, Woosley RL, Oates JA. Tolerance to the humoral and hemodynamic effects of caffeine in man. J Clin Invest. 1981;67:1111–7.
14. Izzo JL, Ghosal A, Kwong T, Freeman RB, Jaenike JR. Age and prior caffeine use alter the cardiovascular and adrenomedullary responses to oral caffeine. Am J Cardiol. 1983;52:769–73.

15. Forman J, Aizer A, Young CR. Myocardial infarction resulting from caffeine overdose in an anorectic woman. Ann Emerg Med. 1997;29:178–80.
16. Bellet S, Horstmann E, Roman LR, DeGuzman NT, Kostis JB. Effect of caffeine on the ventricular fibrillation threshold in normal dogs and dogs with acute myocardial infarction. Am Heart J. 1972;84:215–27.
17. Ishida S, Ito M, Takahashi N, Fujino T, Akimitsu T, Saikawa T. Caffeine induces ventricular tachyarrhythmias possibly due to triggered activity in rabbits in vivo. Jpn Circ J. 1996;60:157–65.
18. Mehta A, Jain AC, Mehta MC, Billie M. Caffeine and cardiac arrhythmias. An experimental study in dogs with review of literature. Acta Cardiol. 1997;52:273–83.
19. Newcombe P, Renton K, Rautaharju P, Spencer A, Montague T. High-dose caffeine and cardiac rate and rhythm in normal subjects. Chest. 1988;94:90–4.
20. Bioh G, Gallagher MM, Prasad U. Survival of a highly dose of caffeine. BMJ Case Rep. 2013. https://doi.org/10.1136/bcr-2012-007454.
21. Chelsky LB, Cutler JE, Griffith K. Caffeine and ventricular arrhythmias. An electrophysiological approach. JAMA. 1990;264:2236–40.
22. Sutherland D, McPherson D, Renton K, Spencer C, Montague T. The effects of caffeine on cardiac rate, rhythm, and ventricular repolarization. Chest. 1985;87:319–24.
23. Harris A, Bluest R. Treatment of slow heart rate following acute myocardial infarction. Br Heart J. 1966;28:631–7.
24. Dobmeyer D, Stine R, Leier C, Greenberg R, Schaal S. The arrhythmogenic effects of caffeine in human beings. N Engl J Med. 1983;308:814–6.
25. Myers M, Harris L, Leenen F, Grant D. Caffeine as a possible cause of ventricular arrhythmias during the healing phase of acute myocardial infarction. Am J Cardiol. 1987;59:1024–8.
26. Lemery R, Pecarskie A, Bernick J, Williams K, Wells G. A prospective controlled randomized study of caffeine in patients with supraventricular tachycardia undergoing electrophysiologic testing. J Cardiovasc Electrophysiol. 2015;26:1–6.
27. Gould L, Reddy CV, Oh KC, Kim SG, Becker WH. Electrophysiologic properties of coffee in man. J Clin Pharmacol. 1979;19:46–55.
28. Rashid A, Hines M, Scherlag BJ, Yamanashi WS, Lovallo W. The effects of caffeine on the inducibility of atrial fibrillation. J Electrocardiol. 2006;39:421–5.
29. Di Rocco JR, During A, Morelli PJ, Heyden M, Biancaniello TA. Atrial fibrillation in healthy adolescents after highly caffeinated beverage consumption: two case reports. J Med Case Reports. 2011;5:18.
30. Frost L, Vestergaard P. Caffeine and risk of atrial fibrillation or flutter: the Danish diet, cancer and health study. Am J Clin Nutr. 2005;81:578–82.

How to Induce Arrhythmias with Adenosine?

14

Simona Căinap

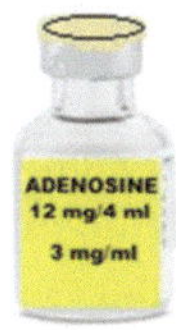

14.1 Introduction

Adenosine is a nucleoside that is normally found inside the human body (Fig. 14.1). It has a negative chronotropic and dromotropic effect. It slows the sinus node rate and also the conduction through the AV node [1].

The use of adenosine in cardiology dates back many years. In 1927 it was used for the first time demonstrating that a cardiac extract containing adenosine can slow the heart rate. After 50 years, in 1980s it was used for termination of supraventricular tachycardias. In 1989 it was approved by FDA for emergency use in paroxysmal SVT [2].

Nowadays it is used in the emergency department for termination of paroxysmal supraventricular tachycardia. It is also useful for the diagnosis of large QRS tachycardia especially when the tachycardia is unresponsive to lidocaine. Adenosine is also a known potent vasodilator of coronary and cerebral resistance vessels. However, despite its effects on termination of reentrant arrhythmias depending on AV node, many studies reported ventricular and supraventricular arrhythmias related to the use of adenosine [3].

S. Căinap (✉)
"Iuliu Hatieganu University of Medicine and Pharmacy", Pediatric Clinic No 2, Pediatric Cardiology, Cluj-Napoca, Romania

G. Cismaru (ed.), *Arrhythmia Induction in the EP Lab*,
https://doi.org/10.1007/978-3-319-92729-9_14

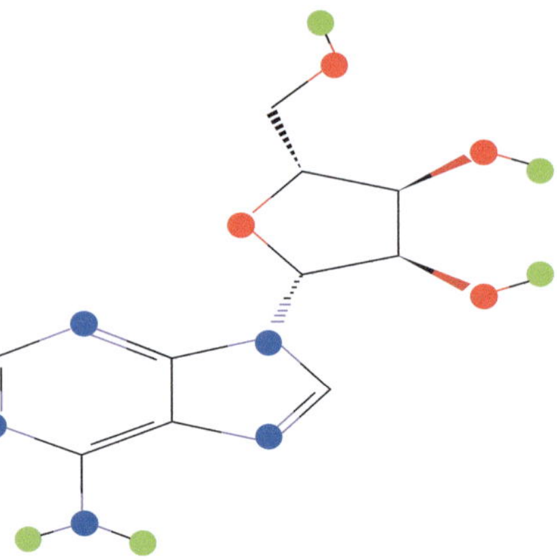

Fig. 14.1 Chemical structure of adenosine

14.2 Pharmacology

Adenosine has a very short life in the human body, the half-life being less than 10 sec. Because of the adenosine receptors that are distributed all across the organs, the effects are extended to different organ systems. Adenosine is the structural base of both ATP and cAMP. Inside the cells, adenosine is produced by degradation of ATP. Adenosine binds to A1, A2A, A2B, and A3 receptors. The effect on A2A and A2B consists of vasodilatation in the coronary circulation. Activation of A1 receptors has a negative chronotropic and dromotropic effect [4].

14.3 Mechanism of Action

Adenosine's mechanism of action is mainly on the sinus node and atrioventricular node. It stimulates the potassium current I_K, Ado and ACh, inducing cellular hyperpolarization and blocking the conduction through the AV node with subsequent interruption of PSVT that is node-dependent.

At the atrial level, adenosine shortens the action potential duration and the refractory periods; thus it can induce supraventricular premature beats and atrial fibrillation. Adenosine stimulates the I_{KAdo} which is the outward potassium current. This increase leads to hyperpolarization of the resting membrane potential of the myocardial cells from the muscular sleeves of the pulmonary veins [5].

Adenosine increases the effect of cAMP by acting on the A2 receptor. Adenosine is antagonized by caffeine and methylamines and potentiated by medication that blocks the nucleoside transport, like dipyridamole.

Bradycardia that is induced by adenosine is very unlikely to respond to atropine.

Digoxin and verapamil have been associated with ventricular fibrillation and are contraindicated in association with adenosine in patients with atrial fibrillation conducted anterogradely through an accessory pathway.

Carbamazepine can increase the degree of heart block produced by adenosine.

14.4 Doses of Adenosine

In case of atrial fibrillation ablation, adenosine should be given as a bolus of 6 mg with increase of 6 mg for successive boli until maximum of 18 mg is reached or atrial ectopy is induced.

For paroxysmal supraventricular tachycardia induction, adenosine is given in small doses, 1–2 mg with continuous atrial stimulation, to obtain prolongation of the AH interval and AVNRT induction (Tables 14.1, 14.2, 14.3, and 14.4).

Table 14.1 Protocol for administration of adenosine

1. The patient should not eat for at least 8 h before the electrophysiological study
2. Patients are instructed to stop medication five times the half time prior to electrophysiological testing unless instructed otherwise
3. **For atrial fibrillation induction**: After negative programmed atrial stimulation adenosine is given in dose of 6 mg with subsequent boluses of 6 mg until maximum dose of 18 mg is reached or atrial fibrillation is induced
4. For AVNRT or AVRT induction, small dose of adenosine are used 1–2 mg bolus
5. For dormant conduction, at least one P wave should be blocked after adenosine or the R-R interval should be >3 s in case of atrial fibrillation
6. In case of caffeine or aminophylline consumers, for atrial fibrillation induction higher doses as 30 mg are necessary

Table 14.2 Adenosine administration for testing pulmonary vein isolation

Time	Dose
Start	
Isolation of the left pulmonary veins	Give 12 mg of adenosine for immediate LPV testing
After 30 min	Give 12 mg of adenosine for late LPV testing
Isolation of the right pulmonary veins	Give 12 mg of adenosine for immediate RPV testing
After 30 min	Give 12 mg of adenosine for late RPV testing

LPV left pulmonary veins, *RPV* right pulmonary veins

Table 14.3 Protocol for adenosine during electrophysiology study for AF induction

Time	Dose
Start	6 mg of adenosine
After 2 min	6 mg of adenosine
After 2 min	6 mf of adenosine
For caffeine or methylxanthine consumers	Supplementary bolus of 12 mg

Table 14.4 Contraindications for adenosine

• Second- or third-degree AV block
• Sick sinus syndrome
• Long QT syndrome
• Severe hypotension
• Decompensated heart failure
• Asthma-relative contraindication

14.5 Preparation

Methylxanthines such as theophylline and caffeine can inhibit the A1 receptor and thus completely block the action of adenosine. Sometimes high doses of adenosine are required in case of theophylline or caffeine users, 30 mg as mentioned above [6].

14.6 Side Effects of Adenosine [7]

Major side effects are uncommon with adenosine. There are published case reports of acute bronchospasm after adenosine administration in patients with asthma or COPD (Tables 14.4 and 14.5).

Table 14.5 Side effects of adenosine

• Facial flushing
• Headache
• Sweating
• Palpitations
• Chest pain
• Hypotension
• Shortness of breath
• Chest pressure
• Hyperventilation
• Head pressure
• Lightheadedness
• Dizziness
• Tingling in the arms
• Numbness
• Apprehension
• Blurred vision
• Burning sensation
• Heaviness in the arms
• Neck and back pain
• Nausea
• Metallic taste
• Tightness in the throat
• Pressure in the groin
• Asystole
• Ventricular tachycardia
• Ventricular fibrillation
• Transient increase in blood pressure
• Sinus bradycardia
• Atrial fibrillation
• Polymorphic ventricular tachycardia
• Torsade des pointes
• Bronchospasm
• Seizures
• Loss of consciousness

Minor side effects like chest pain headache and flushing occur in 10% of the patients [8].

14.7 Electrophysiological Effects of Adenosine

14.7.1 Animal Studies

The first study with adenosine on animals was that of Drury and Szent. They isolated a substance similar to adenosine from the heart muscle and administered it to guinea pigs obtaining complete AV block. Following this study, other observations were made on rabbits, sheeps, pigs, and cats [9].

In 2010 Datino et al. demonstrated on dogs that adenosine administration predicted spontaneous reconnection of the pulmonary veins at 30–60 min after injection [10].

14.7.2 Human Studies

Human studies have shown that adenosine administration reveals dormant PV conduction. The ADVICE trial demonstrated that 53% of the patients that received adenosine had dormant conduction. They were randomized to receive either further ablation or no ablation. The first group showed a reduction of 27.1% in arrhythmia recurrences [11].

The UNDER-ATP trial patients were randomized to receive adenosine or no adenosine 57 min after pulmonary vein isolation. The results showed that more than 40% of patients from both groups presented spontaneous reconnection [12].

14.7.2.1 Adenosine for AVNRT Induction

Adenosine is used in the emergency department to stop supraventricular tachycardia. But it can also induce PSVT.

Curtis et al. demonstrated in a beautiful study that low dose of adenosine can be used to induce AVNRT. The fast and the slow AV node pathways do not have a similar response to adenosine. In case of small doses of adenosine, the antegrade fast pathway is more sensitive than the slow pathway. Adenosine blocks the fast pathway, facilitating conduction through the slow pathway, with induction of AV node reentrant tachycardia. In emergency practice, clinicians use high doses of adenosine that block unselectively the slow and the fast pathway, with cessation of the tachycardia [13]. The mean dose used for AVNRT induction in the study of Curtis was 3.2 ± 2.9 mg.

Tebbenjohanns et al. injected increasing doses of adenosine, 3, 6, 9, 12, 15, and 18 mg in 37 patients with a mean age of 50 years, and demonstrated a sudden jump of >50 ms in patients with dual nodal physiology. In four patients (23%), after the jump, AVNRT was induced by adenosine injection.

14.7.2.2 Adenosine for AVRT Induction

Adenosine can induce reentry using an accessory pathway. By slowing the anterograde conduction through the AV node, it allows retrograde activation of the atrium through the accessory pathway [14, 15]. Garrat et al. reported an increase in the heart rate of a paroxysmal AV reentrant tachycardia after adenosine. It can also increase the ventricular rate in case of AV reentrant tachycardia [16] by modifying the electrical properties of the retrograde limb of the reentrant circuit. Other cases of PSVT acceleration were reported by Orenbaugh et al. [17].

14.7.2.3 Adenosine for PVC Induction

PVCs are frequently present during adenosine administration [18]. Nonsustained VT was also reported after adenosine given for SVT [19].

DiMarco et al. found PVCs in one third of the patients with adenosine administrated for supraventricular tachycardia, the most plausible mechanism being increased automatically [20].

In the study of Tan et al. on 187 episodes of PSVT, during a 5-year period, two thirds of the patients presented premature ventricular complexes or nonsustained ventricular tachycardia. More than 50% of the patients presented PVC with a particular morphology suggesting origin in the inferior left ventricular septum: right bundle branch block with superior axis [21].

One of the mechanisms for PVC induction could be enhanced automatically because adenosine increases the adrenergic tone [22]. Furthermore, delay after depolarizations could also be incriminated because PVCs occur after fast rates [23]. The particular morphology of RBBB and superior axis suggest origin in the Purkinje fibers of the posteroinferior fascicle. These cells are more likely to generate ectopic beats as they have a poor electrical coupling [24].

In our opinion adenosine could be used for PVC ablation only when PVCs are absent at the beginning of the EP study. At least one clinical PVC should be induced in order to search for match during pacemapping. It is mandatory that the induced PVC shows the same morphology with the clinical PVC obtained on a 12-lead ECG.

14.7.2.4 Adenosine for AF Induction

The reported frequency for adenosine-induced atrial fibrillation reaches nearly 12% [25].

In the study of Isa-Param, adenosine and isoproterenol showed a high rate of inducibility of atrial arrhythmias in patients referred for atrial fibrillation ablation (86% and 82%, respectively) [26].

At the atrial level, adenosine shortens the action potential duration and the effective as well as relative refractory period. It also increases the reflex sympathetic response causing arterial hypotension [27, 28]. By modifying the effective and absolute refractory periods, adenosine generates the development of functional reentry [29]. The reentrant activity has a predilection for the pulmonary veins, due to the anisotropic conduction and heterogeneity of repolarization. Despite the fact that adenosine has a negative chronotropic and dromotropic effect, reducing the automatism and pacemaker activity, the increase in the reflex sympathetic response produced by adenosine administration can further induce abnormal automatism and pacemaker activity, after the bradycardia period has ended [30].

Disappearance of atrial fibrillation recurrences after ablation of induced atrial ectopia confirms the pathogenic and clinical importance of adenosine-induced atrial arrhythmias. In the series of Marchlinski et al. [31], ablation guided exclusively by the induced atrial ectopic beats obtained a good long-term success of around 80%. The clinical importance of this finding in our opinion is when ablating extrapulmonary foci. Considering the issues mentioned above, we ablate in our electrophysiological laboratory all the four pulmonary veins, not only those with induced atrial ectopy, as other foci might be left untreated.

Adenosine can mimic the hyperpolarizing effect that occurs spontaneously over time inside the pulmonary veins. Approximately 30 min after catheter ablation, damaged cardiac cells, which are still viable, can show hyperpolarization of the resting potential that lead to reconnection of the isolated pulmonary vein. Adenosine can reveal dormant pulmonary vein conduction after an effective antral isolation. In the multicenter study published by Macle et al., adenosine administration after PV antral isolation, followed by supplemental ablation in case of reconduction, improved procedural outcomes [32].

In a canine study, adenosine administration predicted spontaneous reconnection of the pulmonary veins at 30–60 min after injection [10].

A systematic review published in 2013 showed that adenosine administration after catheter ablation of atrial fibrillation improved the procedural outcomes, especially the freedom from AF [33].

14.7.2.5 Adenosine for VT Induction

Misra et al. described the first case of monomorphic VT after adenosine. The patient had atrial fibrillation and after adenosine presented premature ventricular complexes and ventricular tachycardia with a left bundle branch block morphology. No electrophysiological study was performed to confirm the tachycardia, but the 12-lead ECG criteria were in favor of VT [19].

14.7.2.6 Adenosine for VF Induction

The increase in ventricular automaticity is related to catecholamine release [34]. Furthermore adenosine activates a cAMP-dependent system that sensitizes the myocardium to catecholamine action [35].

Shah et al. reported two cases of ventricular fibrillation in patients with pre-excited atrial fibrillation that received 12 mg adenosine and increased the antegrade conduction to the ventricles through the accessory pathway [36].

Gupta et al. demonstrated in a beautiful study that only accessory with short refractory period is predisposed to ventricular fibrillation after adenosine administration. In patients with long refractory period, atrial fibrillation was induced during electrophysiological study by programmed atrial stimulation, but the shortest RR interval during AF was 300 ms and no VF developed [37].

Rajkumar et al. reported a case of ventricular fibrillation after adenosine administration in a patient without any accessory pathway, which was excluded by an electrophysiological study [38].

Adenosine has been also reported to cause polymorphic ventricular tachycardia in patients with congenital and acquired long QT syndrome, but it is not a marker of LQTS because other authors reported polymorphic VT also in patients with normal QT interval [39].

Parham et al. described degeneration of monomorphic VT into ventricular fibrillation after two boluses of 12 mg of adenosine. The patient had dilated cardiomyopathy and was implanted with a double-chamber pacemaker due to sinus node disease associated with high-degree atrioventricular block [40].

References

1. Honey R, Ritchie W, Thomson W. The action of adenosine upon the human heart. Q J Med. 1930;23:485–9.
2. Lerman BB, Belardinelli L. Cardiac electrophysiology of adenosine. Basic and clinical concepts. Circulation. 1991;83:1499–509.
3. Jaeggi E, Chiu C, Hamilton R, Gilljam T, Gow R. Adenosine induced atrial pro-arrhythmia in children. Can J Cardiol. 1999;15:169–72.
4. Moro C, Lorio N, Nuñez A, Martinez J, Novo L, Aguilera M, et al. Dose related efficacy of adenosine triphosphate in spontaneous supraventricular tachyarrhythmias. Int J Cardiol. 1989;25:207–12.
5. Liu J, Huang CX, Jiang H, Bao MW, Cao F, Wang T. Characteristics of hyperpolarization-activated inward current in rabbit pulmonary vein muscle sleeve cells. Chin Med J. 2005;118:2014–9.
6. Ukena D, Schudt C, Sybrecht GW. Adenosine receptor blocking xanthines as inhibitors of phosphodiesterase isozymes. Biochem Pharmacol. 1993;45:847–51.
7. Hashimoto T, Young WL, Aagaard BD, Joshi S, Ostapkovich ND, Pile-Spellman J. Adenosine-induced ventricular asystole to induce transient profound systemic hypotension in patients undergoing endovascular therapy. Dose-response characteristics. Anesthesiology. 2000;93:998–1001.
8. Gil Madre J, Lázaro Rodriguez S, Sentenac Merchan G, Sepulveda Berocal MA, Alises Moraleda JM, Cortes Bermejo S, et al. Adenosine triphosphate in the treatment of supraventricular paroxysmal tachycardia: a comparison with verapamil. Rev Esp Cardiol. 1995;48:55–8.
9. Drury AN, Szent-Gyorgyi A. The physiological activity of adenine compounds with special reference to their action upon the mammalian heart. J Physiol. 1929;68:213–37.
10. Datino T, Macle L, Qi XY, Maguy A, Comtois P, Chartier D, et al. Mechanisms by which adenosine restores conduction in dormant canine pulmonary veins. Circulation. 2010;121:963–72.
11. Macle L, Khairy P, Verma A, Weerasooriya R, Willems S, Arentz T, et al. Adenosine following pulmonary vein isolation o target dormant conduction elimination (ADVICE): methods and rationale. Can J Cardiol. 2012;28:184–90.
12. Kobori A, Shizuta S, Inoue K, Kaitani K, Morimoto T, Nkazawa Y, et al. Adenosine triphosphate-guided pulmonary vein isolation for atrial fibrillation: the Unmasking Dormant Electrical Reconduction by Adenosine TriPhosphate (UNDER-ATP) trial. Eur Heart J. 2015;36:3276–87.
13. Curtis AB, Belardinelli L, Woodart DA, Brown CS, Conti JB. Induction of atrioventricular node reentrant tachycardia with adenosine: differential effect of adenosine on fast and slow atrioventricular node pathways. J Am Coll Cardiol. 1997;30:1778–84.

14. Dougherty AH, Gilman JK, Wiggins S, et al. Provocation of atrioventricular reentry tachycardia: a paradoxical effect of adenosine. Pacing Clin Electrophysiol. 1993;16:8–12.
15. Vora A, Lokhandwala Y. Adenosine-induced tachycardia: what is the mechanism? J Cardiovasc Electrophysiol. 2000;11:491–2.
16. Garratt CJ, O'Nunain S, Griffith MJ, Connelly DT, Rowland E, Ward DE, et al. Effects of intravenous adenosine in patients with preexcited junctional tachycardias: therapeutic efficacy and incidence of proarrhythmic events. Am J Cardiol. 1994;74:401–4.
17. Orebaugh SL, Handy M. Intravenous adenosine therapy accelerating rate of paroxysmal supraventricular tachycardia. Am J Emerg Med. 1992;10:326–30.
18. Ertan C, Atar I, Gulmez O, Atar A, Ozgul A, Aydinalp A, et al. Adenosine induced ventricular arrhythmias in patients with supraventricular tachycardias. Ann Noninvasive Electrocardiol. 2008;13:386–90.
19. Misra D, Van Tosh A, Schweitzer P. Adenosine induced monomorphic ventricular tachycardia. Pacing Clin Electrophysiol. 2000;23:1044–6.
20. DiMarco JP, Miles W, Akhtar M, Milstein S, Sharma AD, Platia E, et al. Adenosine for paroxysmal supraventricular tachycardia: dose ranging and comparison with verapamil. Assessment in placebo-contrSzent-Guorgyi. Olled, multicenter trials. The Adenosine for PSVT Study Group. Ann Intern Med. 1990;113:104–10.
21. Tan HL, Spekhorst HHM, Peters RJG, Wilde AAM. Adenosine induced ventricular arrhythmias in the emergency room. Pacing Clin Electrophysiol. 2001;24:450–5.
22. Biaggioni I, Killian TJ, Mosqueda-Garcia R, Robertson RM, Robertson D. Adenosine increases sympathetic nerve traffic in humans. Circulation. 1991;83:1668–75.
23. Wit AL, Rosen MR. After depolarizations and triggered activity. In: Fozzard HA, et al., editors. The heart and cardiovascular system. New York: Raven Press; 1986. p. 1449–90.
24. Argentieri TM, Frame LH, Colatsky TJ. Electrical properties of canine subendocardial Purkinje fibers surviving in 1-day-old experimental myocardial infarction. Circ Res. 1990;66:123–34.
25. Strickberger SA, Man KC, Daoud EG, Goyal R, Brinkman K, Knight BP, et al. Adenosine-induced atrial arrhythmia: a prospective analysis. Ann Intern Med. 1997;127:417–22.
26. Isa-Param R, Perez-Castellano N, Villacastin J, Moreno J, Salinas J, Alonso R, et al. Inducibility of atrial arrhythmias after adenosine and isoproterenol infusion in patients referred for atrial fibrillation ablation. Rev Esp Cardiol. 2006;59:559–66.
27. Corbisiero R, Kabell G, Cook JR, Fitzgerald TF, Kirchhoffer JB. Effects of adenosine on local stimulus-response latency and induction of atrial fibrillation by premature stimuli. Pacing Clin Electrophysiol. 1999;22:1378–85.
28. Raviña T, Raviña P, Suárez M. Inestabilidad auricular tras adenosina. Rev Esp Cardiol. 2004;57:594–5.
29. Arora R, Verheule S, Scott L, Navarrete A, Katari V, Wilson E, et al. Arrhythmogenic substrate of the pulmonary veins assessed by high-resolution optical mapping. Circulation. 2003;107:1816–21.
30. Dixit S, Gerstenfeld EP, Callans DJ, Marchlinski FE. Mechanisms underlying sustained firing from pulmonary veins: evidence from pacing maneuvers and pharmacological manipulation. Pacing Clin Electrophysiol. 2004;27:1120–9.
31. Marchlinski F, Callans D, Dixit S, Gerstenfeld E, Rho R, Ren JF, et al. Efficacy and safety of targeted focal ablation versus PV isolation assisted by magnetic electroanatomic mapping. J Cardiovasc Electrophysiol. 2003;14:358–65.
32. Macle L, Khairy P, Weerasooriya R, Novak P, Verma A, Willems S, et al. Adenosine-guided pulmonary vein isolation for the treatment of the paroxysmal atrial fibrillation: an international multicenter, randomized superiority trial. Lancet. 2015;386:672–9.
33. Mc Lellan AJ, Kumar S, Smith C, Morton JB, Kalman JM, Kistler PM. The role of adenosine following pulmonary vein isolation in patients undergoing catheter ablation for atrial fibrillation: a systematic review. J Cardiovasc Electrophysiol. 2013;24:724–51.
34. Hernandez J, Ribeiro JA. Excitatory actions of adenosine on ventricular automaticity. Trends Pharmacol Sci. 1996;17:141–4.

35. Hernandez J, Laorden ML, Ruiz F, Ribeiro JA. Effects of adenosine and its analogues on ventricular automaticity induced by a local injury: role of catecholamines and of cyclic AMP. Arch Int Pharmacodyn Ther. 1989;297:49–59.
36. Shah CP, Gupta AK, Thakur RK, et al. Adenosine induced ventricular fibrillation. Indian Heart J. 2001;53:208–10.
37. Gupta AK, Shah CP, Maheshwari A, Thakur RK, Hayes OW, et al. Adenosine induced ventricular fibrillation in Wolf-Parkinson-White syndrome. Pacing Clin Electrophysiol. 2002;25:477–80.
38. Rajkumar CA, Qureshi N, Siong NGF, Panoulas VF, Lim PB. Adenosine induced ventricular fibrillation in a structurally normal heart: a case report. J Med Case Rep. 2017;11:21.
39. Smith JR, Goldberger JJ, Kadish AH. Adenosine induced polymorphic ventricular tachycardia in adults without structural heart disease. Pacing Clin Electrophysiol. 1997;20:743–5.
40. Parham WA, Mehdirad AA, Biermann KM, Fredman CS. Adenosine induced ventricular fibrillation in a patient with stable ventricular tachycardia. J Interv Card Electrophysiol. 2001;5:71–4.

How to Induce Arrhythmias with Aminophylline?

15

Habib Rehman Khan

15.1 Introduction

Aminophylline, a methylxanthine, is a 2:1 combination between theophylline and ethylenediamine (Fig. 15.1). The 2:1 ratio improves the solubility of the product. Compared to theophylline, aminophylline is a less potent nonselective adenosine receptor antagonist and phosphodiesterase inhibitor [1–3].

Aminophylline is a medication used for the treatment of obstructive lung disease such as asthma and COPD. Other indications for aminophylline use are reversal of adenosine or dipyridamole infusion during nuclear cardiac stress testing, anaphylactic shock (for the bronchodilator effect), and atropine-resistant bradyasystolic arrest.

H. R. Khan (✉)
Royal Brompton and Harefield, NHS Trust, London, UK

National Heart and Lung Institute, Imperial College, London, UK

G. Cismaru (ed.), *Arrhythmia Induction in the EP Lab*,
https://doi.org/10.1007/978-3-319-92729-9_15

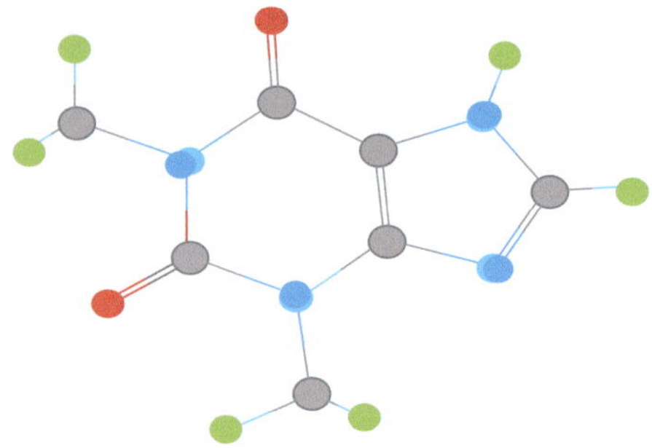

Fig. 15.1 Chemical structure of aminophylline

15.2 Pharmacology

After aminophylline enters in the blood of the patient, 40% of it binds to the plasma proteins, mainly albumin. Approximately 90% of aminophylline is metabolized in the liver.

Nearly 6% of theophylline is metabolized to caffeine. The only metabolites with pharmacologic activity are caffeine and 3-methylxanthine.

A 6 mg/kg dose of aminophylline will produce a serum concentration of 10 μg/mL with a range of 6–16 μg/mL [4]. In geriatric patients, the clearance of theophylline is reduced by 30%, and care should be taken to monitor the occurrence of side effects.

15.3 Mechanism of Action

Aminophylline through its effect on phosphodiesterase inhibitor can cause bronchodilation, diuresis, cardiac stimulation, central nervous system (CNS) stimulation, and gastric acid secretion. Aminophylline also increases the force of contraction of diaphragmatic muscles.

Aminophylline antagonizes adenosine and its effect on the heart. Adenosine is an extracellular compound the regulates oxygen demands in the myocardium. It regulates intercellular pathways through cell surface receptors to increase coronary blood flow, slow heart rate through sinus node, block atrioventricular node (AV) conduction, suppress cardiac automaticity, and decrease β-adrenergic effects on contractility [5, 6]. Adenosine antagonizes effects of circulating catecholamines [2–9].

The effects on the sinus node are reduction in the corrected sinus node recovery time and sinoatrial conduction time. The effect on the atrial tissue is reduction in the atrial refractory period. The effect on the AV node was decrease in the Wenckebach point and AH time. The effect on the His-Purkinje system is reduction in the HV conduction time. There was no change in the ventricular refractory period in the study of Eiriksson et al. [1].

15.4 Protocol for Administration

15.5 Doses of Aminophylline

Aminophylline is available in injection form and oral tablets. Injection concentration is 240 mg/10 mL or 250 mg/10 mL and can be administered as rate-regulated infusion. The rate of infusion should be adjusted until a therapeutic plasma concentrations is achieved (10–20 µg/mL) with toxic effects >20 µg/mL [1, 7, 8]. A usual protocol of administration for arrhythmia induction is 6 mg/kg administered in 30 min with the objective of increasing the heart rate with at least 50% more than the basal resting rate (Tables 15.1, 15.2, and 15.3).

Table 15.1 Protocol for aminophylline administration

1. The patient should not eat or smoke for at least 8 h before the electrophysiological study
2. Patients should not be taking aminophylline or theophylline prior to EP study
3. An IV is started using aseptic technique. Aminophylline can be mixed with NaCl 0.9% or glucose 5%
4. An infusion pump is necessary for aminophylline loading dose and maintenance infusion administration
5. Aminophylline is loaded at a dose of 6 mg/kg over 30 min
6. Continuous monitoring of EKG, heart rate, and 5-min assessment of blood pressures and symptoms for each infusion stage to monitor for toxic effects. Aminophylline has a narrow therapeutic window and therefore risks of toxicity need to be monitored closely
7. When the heart rate increases to more than 50% higher than the resting rate, atrial and ventricular stimulation can be performed for arrhythmia induction

Table 15.2 Side effects of amidophylline

Concentrations <20 µg/mL • Transient caffeine-like adverse effects – Nausea – Vomiting – Headache – Insomnia – Irritability – Restlessness – Transient diuresis – Fine skeletal muscle tremor • Other – Diarrhea – Seizures-rarely
Concentrations >20 µg/mL • Generalized seizures • Persistent vomiting • Abdominal pain • Hematemesis • Hypoglycemia • Hyperglycemia

Table 15.3 How to prepare an aminophylline infusion

Understanding the numbers:

You are ordered to give 6 mg/kg of Aminophylline over 30 min to your 58-year-old, 70 kg patient to facilitate clinical arrhythmia induction

1. The aminophylline vial has 240 mg/10 mL
2. You should take 2 vials of aminophylline in the 50 mL syringe with 30 mL of serum (0.9% NaCl)
3. This gives a concentration of 480 mg/50 mL meaning 9.6 mg/mL
4. 6 mg x 70 kg=420 mg should be injected, equivalent of 44 mL.
5. We will start the perfusion rate: 88 mL/h

15.5.1 Aminophylline Contraindications

1. Acute porphyria
2. Allery to theophyllines

15.5.2 Aminophylline's Side Effects

Gastrointestinal symptoms: Anorexia, nausea, vomiting, diarrhea, and hematemesis

Cardiovascular: Supraventricular and ventricular arrhythmias, hypotension, and palpitations

Neurological: Convulsions, restlessness, insomnia, headache, hallucinations

Lab tests: Theophylline at serum concentrations of 10–20 μg/mL produces moderate increase in plasma glucose, moderate increase in uric acid, moderate increase in free fatty acids, total cholesterol and HDL, and urinary free cortisol excretion. It also produces a moderate decrease in serum T3 (triiodothyronine) (Table 15.2).

Drug interaction: Theophylline blocks adenosine receptors, and higher doses of adenosine are required to stop a supraventricular tachycardia in emergency settings.

15.6 Electrophysiologic Properties

Aminophylline is not routinely used in cardiac electrophysiology testing due to its narrow therapeutic window, and there is no consensus established yet on its use for emergencies such as non-trauma cardiac arrests due to the small number of patients enrolled in studies. The effect of aminophylline on the cardiac myocardium is reduction of sinoatrial conduction time, atrioventricular and His-Purkinje system conduction intervals, and atrial refractory period [5, 6, 9–11].

Several small studies have attempted to assess clinical use of aminophylline for use in cardiac electrophysiology lab but have only been able to establish that it mostly reduces ability to induce arrhythmias in the lab setting if the doses are kept to a therapeutic range and can be helpful in establishing high degrees of AVN block and concealed accessory pathways [12–15].

Aminophylline shortens antegrade and retrograde effective refractory periods of accessory pathway in patients with WPW. Aminophylline can also unmask a concealed accessory pathway, improved conduction in the AV node and His-Purkinje system, and induction of atrial tachycardia [5, 9–14, 16].

To monitor effects of aminophylline, atrial paced drive trains should be tested every 3 min to assess whether the conduction times have increased by 10%. Once the conduction is improved by 10%, the maintenance dose of aminophylline should be maintained at the same dose and full EP testing conduction to assess arrhythmia inducibility. Once arrhythmia is established, the aminophylline infusion can be reduced by 50% every 5 min. End point testing can be reinstated with aminophylline maintenance infusion. No serum levels need to be tested following termination of aminophylline infusion unless side effects persist or there is an unnatural sinus tachycardia response.

In the routine clinical scenarios, interaction with theophylline is usually with asthmatic patients presenting with theophylline toxicity. In toxic theophylline levels (>20 mg/L), there have been reports of ventricular tachycardia, ventricular fibrillation, supraventricular tachycardia (atrioventricular NRT and AVRT), ventricular ectopics, atrial ectopics, atrial fibrillation, and atrial tachycardia [17–19]. Isoprenaline, adrenaline, and noradrenaline should be avoided as they are known to be alkali labile.

15.6.1 Animal Studies with Aminophylline

The effects of aminophylline on canine myocardium were studied by Komadina et al. and Belhassen et al. which showed that conduction through the AV node and His-Purkinje system is enhanced. In the first study, aminophylline decreases the AH interval, the Wenckebach cycle length, and ventricular refractory period. Association of metaproterenol to aminophylline produced greater reduction in the AH interval, HV interval, WCL, and ventricular refractory period [20, 21]. In the second study, Belhassen et al. showed that aminophylline can antagonize the effects of adenosine on the canine myocardium, but this effect is modulated by the autonomic nervous tone: autonomic blockade is required for the antagonization of the electrophysiological actions of adenosine [20].

15.6.2 Human Studies with Aminophylline

Electrophysiological effects of aminophylline were studied in ten male patients with COPD. Drug infusion produced plasma concentration of 15.6 μg/mL and subsequent increase in plasma catecholamine concentration: epinephrine and norepinephrine. The effect on the sinus node was reduction in the corrected sinus node recovery time and sinoatrial conduction time. The effect on the atrial tissue was reduction in the atrial refractory period. The effect on the AV node was decrease in the Wenckebach point and AH time. The effect on the His-Purkinje system was reduction in the HV conduction time. There was no change in the ventricular refractory period [16].

In patients with heart transplantation, the effects of aminophylline on the sinus node are decrease in the sinus node recovery time and corrected sinus node recovery time and decrease in the sinus node cycle length. Therefore administration of theophylline in heart transplant recipients may decrease the risk of bradycardia [22, 23]. Bertolet et al. also demonstrated that theophylline reduces the need of pacemaker implantation after heart transplantation [24].

High doses of theophylline can induce heart rhythm troubles. In case of serum concentrations higher than 30 mg/L, patients develop symptoms and signs of theophylline toxicity. Sessler et al. studied 16 patients with theophylline toxicity and recorded several arrhythmias: atrial fibrillation in 1 patient, supraventricular ectopic beats in 7 patients, runs of atrial ectopic bets in 4 patients, multifocal atrial tachycardia in 1 patient, ventricular premature beats in 13 patients, and runs of ventricular premature beats in 4 patients. In one patient sustained ventricular tachycardia appeared at doses of 66 mg/L [17]. Paloucek et al. in a cumulative review of 320 patients with theophylline toxicity described atrial fibrillation in 2% and paroxysmal supraventricular tachycardia in 9% of patients. Ventricular tachycardia and ventricular fibrillation was noted in 5% of patients with iatrogenic toxicity [25].

The dose that is used in the EP laboratory for arrhythmia induction is also used as loading dose in patients with status asthmaticus. Patients hospitalized in intensive care unit for status asthmaticus receive a loading dose of 6 mg/kg in 30 min, followed by a maintenance dose of 1–1.2 mg/kg/h [26].

15.6.2.1 Aminophylline for Inappropriate Sinus Tachycardia Induction

Marrouche et al. performed catheter ablation in 39 patients with inappropriate sinus tachycardia using a three-dimensional mapping system. Arrhythmia induction was obtained after isoprenaline infusion or in case of failure, after aminophylline administration. The dose used in this study was 6 mg/kg aminophylline administered during 20 min followed by a maintenance dose of 0.5 mg/kg/min. Of the 39 patients, 35 were inducible with isoprenaline, and in 5 patients additional aminophylline was necessary. Catheter ablation was effective in all 39 patients [27].

15.6.2.2 Aminophylline for Atrioventricular Reentrant Tachycardia

Aminophylline shortens the refractory period of the accessory pathway both in antegrade and retrograde direction and facilitates induction of reentrant arrhythmia [11]. Telichowski et al. demonstrated that aminophylline increases the risk of reentrant tachycardia using an accessory pathway in patients with WPW syndrome [28].

Ichikawa et al. reported a case of orthodromic atrioventricular reentrant tachycardia using a right free wall accessory pathway in a patient treated with theophylline that had a serum concentration of 38.5 μg/mL (for a normal therapeutic range of 10–20 μg/mL). Hemodiafiltration was used for theophylline removal, and catheter ablation of the right free wall was successfully performed.

15.6.2.3 Aminophylline for Atrial Fibrillation Induction

Aminophylline shortens the refractory period of the right and left atrium and disperses the recovery of excitability, facilitating atrial fibrillation induction. Varriale et al. reported three cases of rapid onset atrial fibrillation with rapid ventricular rate after intravenous aminophylline 25–30 mg/h. At the moment of AF onset, serum theophylline concentration was 15 mg/dL, 16 mg/dL, and 30 mg/dL, respectively, in the three patients. After discontinuation of aminophylline, atrial fibrillation stopped in the three patients but gin one patient aminophylline was resumed with recurrence of paroxysmal atrial fibrillation. In that single patient, sinus rhythm was restored again after the drug was stopped again [29].

15.6.2.4 Aminophylline for Ventricular Tachycardia Induction

The EHRA/HRS guidelines expert consensus on ventricular tachycardia ablation mention aminophylline for VT induction in case of triggered activity or automaticity [30]. Triggered activity appears from oscillations in the membrane potential during or following an action potential. Automaticity can be the cause of focal VT which is provoked by adrenergic stimulation not related to triggered activity. Aminophylline has an effect in triggered activity VT. The guideline indication is based on an old study from 1990s by Lerman et al. that demonstrated induction of VT in patients pre-treated with dipyridamole. They administered 2.8 mg/kg aminophylline for 3–10 min. Dipyridamole stopped ventricular tachycardia, but aminophylline infusion resumed VT, reversing the effects of dipyridamole. The study of Lerman et al. demonstrated a good effect of dipyridamole in stopping VT associated with triggered activity, and aminophylline with its effect on the adenosine receptor completely reversed the effects of dipyridamole. Aminophylline can also show inhibition of phosphodiesterase and norepinephrine release from nerve terminations, but these effects occur in a lesser percentage compared to antagonization of adenosine A1 receptor [6, 10, 31].

References

1. Abu-Laban RB, et al. Aminophylline in bradyasystolic cardiac arrest: a randomised placebo-controlled trial. Lancet. 2006;367(9522):1577–84.
2. Hayward E, Showler L, Soar J. Aminophylline in bradyasystolic cardiac arrest. Emerg Med J. 2007;24(8):582–3.
3. Hurley KF, Magee K, Green R. Aminophylline for bradyasystolic cardiac arrest in adults. Cochrane Database Syst Rev. 2013;(8):CD006781.

4. Stirt JA, Sullivan SF. Aminophylline. Anesth Analg. 1981;60(8):587–602.
5. Ostapiuk M, et al. [Effect of aminophylline on electrophysiological characteristics of the accessory pathway in patients with Wolff-Parkinson-White syndrome]. Kardiol Pol. 1988;31(3):173–9.
6. Lerman BB, et al. Antiadrenergic effects of adenosine on His-Purkinje automaticity. Evidence for accentuated antagonism. J Clin Invest. 1988;82(6):2127–35.
7. Prokocimer PG, et al. Epinephrine arrhythmogenicity is enhanced by acute, but not chronic, aminophylline administration during halothane anesthesia in dogs. Anesthesiology. 1986;65(1):13–8.
8. Stirt JA, et al. Aminophylline pharmacokinetics and cardiorespiratory effects during halothane anesthesia in experimental animals. Anesth Analg. 1980;59(3):186–91.
9. Molenda W, et al. [Effect of aminophylline on the heart conduction system in patients with paroxysmal nodal tachycardia]. Kardiol Pol. 1988;31(2):84–90.
10. Molenda W, et al. [Effect of aminophylline on the ventricular part of the heart conduction system]. Pol Arch Med Wewn. 1988;79(5):256–62.
11. Molenda W, et al. Effect of aminophylline on the heart conduction system in patients with concealed accessory atrioventricular node bypass. Pol Tyg Lek. 1987;42(10):287–90.
12. Lerman BB, Wesley RC, Belardinelli L. Electrophysiologic effects of dipyridamole on atrioventricular nodal conduction and supraventricular tachycardia. Role of endogenous adenosine. Circulation. 1989;80(6):1536–43.
13. Malcolm AD, Garratt CJ, Camm AJ. The therapeutic and diagnostic cardiac electrophysiological uses of adenosine. Cardiovasc Drugs Ther. 1993;7(1):139–47.
14. Montenero AS, et al. [Electrophysiologic and hemodynamic effects of intravenous aminophylline in humans]. Cardiologia. 1987;32(9):1005–7.
15. Nicholson DP, Chick TW. A re-evaluation of parenteral aminophylline. Am Rev Respir Dis. 1973;108(2):241–7.
16. Eiriksson CE Jr, Writer SL, Vestal RE. Theophylline-induced alterations in cardiac electrophysiology in patients with chronic obstructive pulmonary disease. Am Rev Respir Dis. 1987;135(2):322–6.
17. Sessler CN, Cohen MD. Cardiac arrhythmias during theophylline toxicity. A prospective continuous electrocardiographic study. Chest. 1990;98(3):672–8.
18. Ichikawa K, et al. A case of life-threatening supraventricular tachycardia storm associated with theophylline toxicity. J Cardiol Cases. 2017;15(4):125–8.
19. Levine JH, Michael JR, Guarnieri T. Multifocal atrial tachycardia: a toxic effect of theophylline. Lancet. 1985;1(8419):12–4.
20. Belhassen B, et al. Autonomic modulation of aminophylline influence on the electrophysiological effects of adenosine and adenosine triphosphate in the canine heart. Cardiovasc Res. 1987;21(2):151–60.
21. Komadina KH, et al. Electrophysiologic study of the effects of aminophylline and metaproterenol on canine myocardium. Chest. 1992;101(1):232–8.
22. Redmond JM, et al. Use of theophylline for treatment of prolonged sinus node dysfunction in human orthotopic heart transplantation. J Heart Lung Transplant. 1993;12(1 Pt 1):133–8; discussion 138–9.
23. Heinz G. Bradycardia after heart transplantation: reversal with theophylline. J Am Coll Cardiol. 1997;29(2):470–1.
24. Bertolet BD, et al. Bradycardia after heart transplantation: reversal with theophylline. J Am Coll Cardiol. 1996;28(2):396–9.
25. Paloucek FP, Rodvold KA. Evaluation of theophylline overdoses and toxicities. Ann Emerg Med. 1988;17(2):135–44.
26. Laaban JP, et al. Cardiac arrhythmias during the combined use of intravenous aminophylline and terbutaline in status asthmaticus. Chest. 1988;94(3):496–502.
27. Marrouche NF, et al. Three-dimensional nonfluoroscopic mapping and ablation of inappropriate sinus tachycardia. Procedural strategies and long-term outcome. J Am Coll Cardiol. 2002;39(6):1046–54.

28. Telichowski C, et al. [Effect of aminophylline on the occurrence of reciprocal tachycardia in patients with Wolff-Parkinson-White syndrome]. Wiad Lek. 1987;40(19):1353–7.
29. Varriale P, Ramaprasad S. Aminophylline induced atrial fibrillation. Pacing Clin Electrophysiol. 1993;16(10):1953–5.
30. Aliot EM, et al. EHRA/HRS Expert Consensus on Catheter Ablation of Ventricular Arrhythmias: developed in a partnership with the European Heart Rhythm Association (EHRA), a Registered Branch of the European Society of Cardiology (ESC), and the Heart Rhythm Society (HRS); in collaboration with the American College of Cardiology (ACC) and the American Heart Association (AHA). Heart Rhythm. 2009;6(6):886–933.
31. Lerman BB. Response of nonreentrant catecholamine-mediated ventricular tachycardia to endogenous adenosine and acetylcholine. Evidence for myocardial receptor-mediated effects. Circulation. 1993;87(2):382–90.

16 How to Induce Arrhythmias with Ephedrine?

Arash Moosavi-Shalheh

16.1 Introduction

Ephedrine is an alpha-beta agonist widely used for nasal decongestion, nervous system stimulation, and spinal anesthesia during cesarean section (Fig. 16.1). Ephedra was used more than 5000 years ago in China and India for treating fever, flu, cold, chills, nasal congestion, coughing, and wheezing. Pseudoephedrine is an isomer of ephedrine [1].

Recently, ephedrine or other herbal products containing ephedra have been used to enhance sport performance and to lose weight [2]. However, ephedrine is prohibited by the International Olympic Committee and the World Anti-Doping

A. Moosavi-Shalheh (✉)
Bispmotalagatan 2 lgh 1301, Motala, Sweden

G. Cismaru (ed.), *Arrhythmia Induction in the EP Lab*,
https://doi.org/10.1007/978-3-319-92729-9_16

Fig. 16.1 Chemical structure of ephedrine: $C_{10}H_{15}NO$

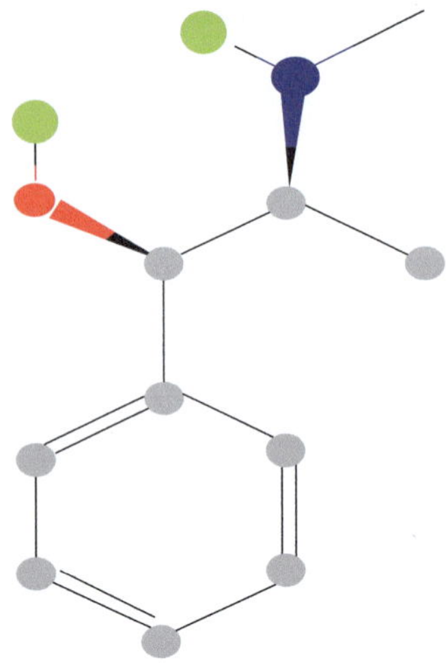

Agency [3, 4] because many studies reported ventricular and supraventricular arrhythmias related to the use of ephedrine [5].

16.2 Pharmacology

When given intramuscular or subcutaneous, ephedrine is rapidly absorbed. The onset of action after muscular administration is between 10 and 20 min. The increase in heart rate after intravenous administration is between 1 and 5 min. The duration of increased heart rate and blood pressure after 10–25 mg of intravenous ephedrine is 1 h and after subcutaneous or intramuscular administration of 25–50 mg is also 1 h.

The metabolization of ephedrine takes place in the liver in a small percentage, because most of it is excreted unmetabolized in the urine. The plasma half-life of ephedrine is 2–3 h [6].

Ephedrine can cross the placenta and can be excreted into the breast milk.

Ephedrine is extremely toxic at doses of 5–50 mg/kg and is lethal in doses higher than 50 mg/kg [7].

16.3 Mechanism of Action

Ephedrine is a sympathomimetic drug that acts both on α- and β-adrenergic receptors. It directly stimulates α_1, β_1, and β_2 adrenergic receptors. Furthermore, it indirectly stimulates the release of noradrenaline from the storage sites. The most important effects are the increase in the heart rate, increase in systolic and diastolic blood pressure, peripheral vasoconstriction, and relaxation of smooth muscle cells of bronchi.

By adrenergic stimulation, ephedrine shortens atrial and ventricular refractory periods, permitting the development of both automatic and reentrant arrhythmias.

It is a drug that is used in obstetrics to treat hypotension that is associated with spinal anesthesia [8]. Ephedrine also has stimulant effects at the level of central nervous system.

16.4 Doses of Ephedrine

The intravenous adult dose should not exceed 150 mg per 24 h [9] (Tables 16.1 and 16.2). Tachyphylaxis can occur after increased ephedrine doses due to depletion of noradrenaline stores.

16.5 Infusion Preparation

Ephedrine is contraindicated in patients with asymmetric septal hypertrophy or idiopathic hypertrophic subaortic stenosis because it improves the myocardial contractility and increases the left ventricular outflow tract obstruction.

Table 16.1 Protocol for ephedrine administration

1. The patient should not eat for at least 8 h before the electrophysiological study
2. Patients are instructed to stop medication five times the half time prior to electrophysiological testing unless instructed otherwise
3. Ephedrine is compatible with 0.9% sodium chloride, 10% glucose, and with Ringer lactate solution
4. It can be administered either as a bolus: first bolus 0.5 mg/kg followed by boluses of 20 mg every 5 min: until sinus tachycardia 150 bpm is obtained or side effects occur
5. Or it can be given as an infusion: 2 ampoules/vials of 50 mg/mL can be mixed with 48 mL of 10% glucose or saline solution with an infusion rate of 50 mL/h
6. Ephedrine injection can result in tissue necrosis and sloughing in case of extravasation
7. In case of important cardiac side effects: high blood pressure >200/100 mmHg or sinus tachycardia >180 bpm. Discontinuation of ephedrine should be initiated with administration of intravenous β blockers: (like betaloc or metoprolol 5 mg) to decrease heart rate and blood pressure

Table 16.2 Ephedrine administration during EP study

Time	Dose
Start	0.5 mg/kg bolus
After 5 min	20 mg bolus
After 5 min	20 mg bolus
After 5 min	20 mg bolus
Do not exceed 150 mg ephedrine in total	

Table 16.3 Contraindications for ephedrine

Contraindications
• Anxiety
• Restlessness
• High blood pressure > 180/100 mmHg
• Asymmetric septa 1 hypertrophy (subaortic stenosis)
• Closed-angle glaucoma
• Prostate adenoma
• Pheochromocytoma
• Thyrotoxicosis
• Unstable ischemic heart disease
• Treatment with monoamine oxidase inhibitors

16.6 Contraindications to Ephedrine

When administered with MAOIs, or within 14 days after stopping the therapy with MAOIs, ephedrine's cardiac effects are prolonged and more intense.

Minor side effects like chest pain and headache can occur, but they rapidly disappear on discontinuation of ephedrine.

In association with cyclopropane or halothane, ephedrine is contraindicated because anesthesia increases cardiac irritability and severe and resistant ventricular arrhythmias can occur.

In geriatric males, especially those with prostatic hypertrophy, ephedrine should be administered with caution because of the risk of acute urinary retention.

In patients with unstable ischemic heart disease, ephedrine should be avoided as it increases oxygen demands in ischemic myocardium (Table 16.3).

16.7 Side Effects of Ephedrine

The principal concern of FDA that banned dietary supplements containing ephedrine was the reported side effects: acute myocardial infarction, hemorrhagic and ischemic stroke, and coronary vasospasm.

Other reported side effects of ephedrine are nausea, anxiety, insomnia, diarrhea, and personality change.

Gedevanishvili et al. reported a case of acute coronary syndrome after 10 mg intravenous ephedrine in a 26-year-old pregnant woman with normal coronary arteries. Furthermore, she developed 10 beats of nonsustained ventricular tachycardia.

During pregnancy, high doses of ephedrine can cause maternal tachycardia and fetal acidosis [10]. Natarajan et al. compared ephedrine 6 mg with phenylephrine 100 μg for hypotension during spinal anesthesia for cesarean section and found that they are equally efficient without differences in terms of fetal acidosis. Only maternal bradycardia was more present in the phenylephrine group [11] (Table 16.4).

Table 16.4 Side effects of ephedrine

Side effects
• Pallor
• Fever
• Headache
• Dryness of nose, mouth, throat
• Angina
• Hypertension
• Nausea, vomiting, epigastric distress
• Anxiety, restlessness
• Insomnia, mood, mental changes
• Fear, irritability, trembling
• Dizziness, lightheadedness,
• Vertigo, confusion, delirium, euphoria
• Psychosis (paranoia, hallucinations, depression after large doses)
• Acute urinary retention
• Dyspnea
• Sweating

16.8 Electrophysiological Effects of Ephedrine

16.8.1 Animal Studies

Adamson et al. verified the hypothesis that ephedrine could increase the risk of fatal arrhythmias on dogs. Fifteen dogs with myocardial infarction received ephedrine, and 9 of the 15 dogs presented an increased percentage of ventricular arrhythmias. Of the nine dogs, four presented ventricular fibrillation, and three could not be resuscitated. After drug washout, none of the surviving animals presented ventricular arrhythmias. Serum catecholamines were measured, and norepinephrine increased four times after ephedrine administration (from 82 to 345 pg/mL) [12].

16.8.2 Human Studies

For maternal hypotension prevention, Loughrey et al. compared ephedrine with ephedrine plus phenylephrine.

They found no difference between the two groups in terms of hypotension, maternal side effects, and fetal acidosis [13].

16.8.2.1 Doses of Ephedrine

There is no study or guideline that recommends ephedrine for arrhythmia induction in the EP laboratory. The doses that we use are extrapolated from studies with ephedrine in pregnant women.

Doses of 10–20 or 30 mg of ephedrine or 0.25 mg/kg were not effective in several studies in eliminating spinal-related hypotension during cesarean section [14–16].

Ngan Kee et al. used fixed dose 30 mg ephedrine as a bolus to prevent hypotension during spinal anesthesia for cesarean section [17]. Datta et al. compared fixed dose of 30 mg with multiple boluses of 10 mg withheld until the moment when hypotension occurred, and it was found that a single bolus of 30 mg was more effective than multiple boluses of 10 mg [18].

Ozdemir et al. demonstrated that a fixed dose of 0.5 mg/kg intravenous ephedrine is effective for preventing hypotension during spinal anesthesia performed for cesarean delivery [19].

We usually do not exceed 150 mg ephedrine for arrhythmia induction during an electrophysiological study. If the dose we used before catheter ablation was 150 mg, then at the end of the procedure, we verify the success of ablation using the same dose of 150 mg.

In case of ephedrine overdose, studies have shown that a contraction band necrosis appears. This type of necrosis is frequently seen in cases of catecholamine excess [20] and was also demonstrated after ephedrine using endomyocardial biopsy guided by three-dimensional electroanatomical mapping.

The minimum lethal dose in mammals is 100–300 mg/kg.

16.8.2.2 Ephedrine for VT Induction

In our EP lab, we use ephedrine for VT induction in patients with idiopathic VT like RVOT, LVOT, and fascicular VT when the tachycardia is noninducible at the beginning of the procedure. Our protocol starts with basal stimulation, and in case of noninduction from the RVOT and RV apex, we inject ephedrine 0.5 mg/kg and then boluses of 20 mg until VT is induced or maximum dose of 150 mg is reached.

Rakovec et al. reported a case of fascicular VT in a 19-year-old woman who used ephedrine to enhance her performance in sports. VT was resistant to both amiodarone and electrical cardioversion and stopped suddenly after 60 h probably after elimination of ephedrine from the patient's plasma.

Casella et al. reported two cases of ventricular tachycardia in two competitive athletes. The VT morphologies were suggestive of RVOT VT. The place of primodepolarization was identified by activation mapping, and RF ablation could make the VT uninducible. Electroanatomical mapping in both patients showed low-voltage areas at the level of RVOT. CARTO-guided endocardial biopsy showed contraction band necrosis, consistent with overstimulation of the adrenergic system. This type of myocardial damage is also observed in cases with excess catecholamines such as pheochromocytoma or cocaine abuse [21].

References

1. Henry B. Pseudoephedrine dilemma. W V Med J. 2014;110:6.
2. Rodos C. Ephedra ban: no shortage of reasons. FDA Consum. 2004;38:6–7.
3. Maughan RJ. Contamination of dietary supplements and positive drug tests in sport. J Sports Sci. 2005;23:883–9.
4. Furlanello F, Bentivegna S, Cappato R, De Ambroggi L. Arrhythmogenic effects of illicit drugs in athletes. Ital Heart J. 2003;4:829–37.

5. Gajraj NM, Wallace DH, Pace NA. Supraventricular tachycardia in a parturient under spinal anesthesia. Reg Anesth. 1993;18:261–3.
6. Thurn AL, Betz JM. Ephedra. In: Coates PM, Betz JM, Blackman MR, et al., editors. Encyclopedia of dietary supplements. 2nd ed. London, UK: Informa Healthcare; 2010. p. 250–6.
7. Gosselin RE, Smith RP, Hodge HC. Clinical toxicology of commercial products. 5th ed. Baltimore: Williams and Wilkins; 1984. p. II–370.
8. Stoelting RK. Sympathomimetics. In: Stoelting RK, editor. Pharmacology and physiology in anesthetic practice. 3rd ed. Philadelphia: Lippincot-Raven; 1999. p. 259–77.
9. Schier JG, Traub SJ, Hoffman RS, Nelson LS. Ephedrine-induced cardiac ischemia: exposure confirmed with a serum level. J Toxicol Clin Toxicol. 2003;41:849–53.
10. Cooper DW, Carpenter M, Mowbray P, Desira WR, Ryall DM, Kokri MS. Fetal and maternal effects of phenylephrine and ephedrine during spinal anesthesia for cesarean delivery. Anesthesiology. 2002;97:1582–90.
11. Natarajan AK, Singh NR, Singh LP, Devi RS, Devi NA, Jack A. Comparison of intravenous bolus phenylephrine and intravenous ephedrine during crystalloid in ameliorating hypotension under spinal anesthesia for cesarean section. J Med Soc. 2015;29:155–9.
12. Adamson PB, Suarez J, Ellis E, Kanaly T, Vanoli E. Ephedrine increases ventricular arrhythmias in conscious dogs after myocardial infarction. J Am Coll Cardiol. 2004;44:1675–8.
13. Loughrey JP, Yao N, Datta S, Segal S, Pian-Smith M, Tsen LC. Hemodynamic effects of spinal anesthesia and simultaneous intravenous bolus of combined phenylephrine and ephedrine versus ephedrine for cesarean delivery. Int J Obstet Anesth. 2005;14:43–7.
14. Chan WS, Irwin MG, Tong WN, Lam YH. Prevention of hypotension during spinal anaesthesia for cesarean section: ephedrine infusion versus fluid preload. Anaesthesia. 1997;52:908–13.
15. King SW, Rosen MA. Prophylactic ephedrine and hypotension associated with spinal anesthesia for cesarean delivery. Int J Obstet Anesth. 1998;7:18–22.
16. Lee A, Ngan Kee WD, Gin T. A dose-response meta- analysis of prophylactic intravenous ephedrine for the prevention of hypotension during spinal anesthesia for elective cesarean delivery. Anesth Analg. 2004;98:483–90.
17. Ngan Kee WD, Khaw KS, Lee BB, Lau TK, Gin T. A dose-response study of prophylactic intravenous ephedrine for the prevention of hypotension during spinal anesthesia for cesarean delivery. Anesth Analg. 2000;90:1390–5.
18. Datta S, Alper MH, Osthemer GW, Weiss JB. Method of ephedrine administration and nausea and hypotension during spinal anesthesia for cesarean section. Anesthesiology. 1982;56:68–70.
19. Kol IO, Kaygusuz K, Gursoy S, Cetin A, Kahramanoglu Z, Ozkan F, et al. The effects of intravenous ephedrine during spinal anesthesia for cesarean delivery: a randomized controlled trial. J Korean Med Sci. 2009;24:883–8.
20. Cassela M, Dello Russo A, Izzo G, Pieroni M, Andreini D, Russo E, et al. Ventricular arrhythmias induced by long-term use of ephedrine in two competitive athletes. Heart Vessels. 2015;30:280–3.
21. Abraham J, Mudd JO, Kapur NK, Champion HC, Klein K, Wittstein IS. Stress cardiomyopathy after intravenous administration of cathecolamines and beta-receptor agonists. J Am Coll Cardiol. 2009;53:1320–5.

17 Arrhythmias Induction and Energy Drinks

Keith Andrew Chan

17.1 Introduction

Energy drinks have become a widely popular choice among consumers and are marketed as caffeinated beverages that enhance attention, athletic performance, and short-term cognition [1]. These drinks, with their high-sugar mixtures and additives promising performance-enhancing capabilities packaged in a palatable formula, have seen tremendous growth in the dietary supplement industry, particularly aimed at the young adult population [2]. A recent comprehensive assessment on the impact and growth of the energy drink industry in the United States since 2002 has demonstrated a 500% growth rate over 5 years with such beverages accounting for over half of the functional beverage market in America [3]. However, shortly after their introduction, reports of highly deleterious effects of these beverages on the cardiovascular system including sudden cardiac death swept the media in several high-profile cases, allegedly in connection with consumption of large amounts of energy drinks [4]. This chapter intends to analyze the scientific basis of such claims, to analyze the various components of energy drinks and to determine the preponderance of evidence demonstrating the arrhythmogenic effects of energy drinks.

17.2 Components of Beverages

Energy drinks contain a number of additives and supplements in addition to the high concentrations of sugar and carbonate blended into the beverages to give them their characteristic sweet soda flavor. These additives form a sort of "functional beverage

K. A. Chan (✉)
Section of Adult Cardiology, Chong Hua Heart Institute, Chong Hua Hospital, Cebu, Philippines

G. Cismaru (ed.), *Arrhythmia Induction in the EP Lab*,
https://doi.org/10.1007/978-3-319-92729-9_17

cocktail" with various manufacturer claims ranging from the ability to improve endurance to wound healing, appetite suppression, and the improvement of sexual performance [5]. The most common additives found in commercially available energy drinks are summarized in Table 17.1.

17.2.1 Caffeine

In one form or another, remains the most commonly infused additive to energy drinks [6]. Listed in Table 17.2 are the caffeine contents of several popular energy drinks noted at the time of this writing [5]. Caffeine and, rather, its effects on causing arrhythmias have been well documented in Chap. 13. Of note, several commonly included additives in energy drinks, particularly guarana and yerba mate, produce their effects by means of them being naturally rich sources of caffeine and its derivatives [7, 8]. Among the other commonly noted additives, several of them have potentially arrhythmogenic properties of interest noted to be separate from their arrhythmogenic propensities secondary to ED's caffeine content. These include taurine, guarana, ginseng, and yerba mate.

17.2.2 Taurine

Is a sulfur-containing amino acid found in high amounts in most diet regimens around the globe. It is involved in a wide variety of physiologic biochemical processes such as anti-inflammatory reactions, neuromodulation, antioxidation, and the maintenance of normal mitochondrial activity which are mostly facilitated by covalent reactions between taurine and other biological compounds [9]. There are several properties of taurine that made it a substance of interest in the investigation of it being a heavy contributor to energy drink-induced cardiac arrhythmias. It is of interest to note that humans do not synthesize taurine in the liver with normal taurine concentrations in the body being extremely low [10]. Acute large doses of taurine in humans, however, did not result in any particular symptoms of toxicity—this

Table 17.1 Additives in energy drinks

Common functional additives in commercial energy drinks [3, 6]
Caffeine
Taurine
Guarana
Ginseng
Yerba mate
B vitamins
Glucuronolactone
L-carnitine

Table 17.2 Caffeine content in energy drinks

Caffeine content in popular energy drinks [5]	
Brand	Caffeine content per serving (mg)
Diet Rockstar Energy Drink™	80
Full Throttle™	72
Go Girl Sugar Free™	150
Lo-Carb Monster XXL™	80
Monster Energy Assault™	80
Monster Energy XXL™	80
Red Bull Sugar Free™	80
Red Bull™	80
Rockstar Energy Drink™	80
Rockstar Juiced™	80
Wired 294 Caffeine™	147

was attributed to the sodium-dependent taurine transporter present on the cell membrane which acted in an extremely slow manner, one too slow to accommodate the influx of taurine to result in increased intracellular concentrations [10]. However, chronic consumption of taurine has been proposed to bypass these mechanisms by means of either genetic modulation or disruption of the cytosol-mitochondrial equilibrium of taurine with proposed end effect being disruption of membrane enzymes that may alter cell structure and function [11–13]. All of these effects were examined in an extensive review article by Schaffer et al. in 2014 with the main area of concerns being the potential arrhythmogenic effect of taurine in relation to high caffeine intake as well as in the setting of ischemic heart disease [14].

17.2.3 Herbal supplements and other components

The effects of herbal additives in energy drinks, such as guarana, ginseng, and yerba mate, particularly their role in arrhythmogenesis, are much less elucidated in literature. Guarana seed extract, as stated earlier, contains large amounts of caffeine and are often included in energy drinks for this purpose. Studies on the alleged arrhythmias originating from guarana seed extract-infused drinks have mostly been attributed to its caffeine components [15]. Like guarana seed extracts, yerba mate exerts its psychological and physiologic effects by virtue of its caffeine components. However, analysis on the effects of chronic and high-dose consumption of yerba mate among rats has proven it to be generally safe and well tolerated [16]. Ginseng (*P. ginseng*) is a medicinal herb that has been used for several centuries in Asian countries and boasts a variety of beneficial effects. Its specific components that confer benefits such as insulin sensitization, blood pressure control, and lipid metabolism still remain an area of intensive scientific research [17]. However, despite anecdotal claims of tachyarrhythmias with ginseng, scientific evidence has been less convincing. In the largest recent systematic review conducted by Lee and

Son in 2011, ginseng was found to be largely safe with the authors citing a generally low incidence of side effects which included gastric upset, hypoglycemia, insomnia, and headache and chest discomfort. No adverse cardiac events were reported in the 57 pooled trials and 3471 participants [18].

17.3 Energy Drink Arrhythmias – A review of published cases

Given the components of energy drinks and their theoretical safety and toxicity profiles, it's appropriate that we examine the literature detailing case reports and case series purporting a relation between energy drink consumption and arrhythmias.

The deleterious effects of energy drinks on the cardiovascular system have been extensively published. In a randomized study by Grasser et al. in 2014, subjects that consumed energy drinks were noted to have increased systolic and diastolic blood pressures, increased respiratory rates, increased cerebrovascular resistance, and decreased cerebral blood flow velocity, making for an overall highly negative cardiovascular effect profile post-energy drink consumption [19]. This, plus a combination of the drinks' high-sugar content and frequent consumer behavior of the target demographic, naturally brings concern for more serious cardiovascular events.

An extensive systematic review by Goldfarb et al. in 2014 investigated case reports and incidents involving cardiovascular events reported in relation to heavy consumption of energy drinks or co-consumptions with other additives (i.e., alcohol and drugs). Over a span of 33 years (1980–2013), Goldfarb summarized a total of 17 case reports with 11 cases of cardiac arrhythmia as the initial presentation. Such arrhythmias included atrial fibrillation, supraventricular tachycardia, torsades de pointes, ventricular tachycardia, and ventricular fibrillation. Other presentations described included prolongation of the QT interval in one patient and cardiac arrest in another report [4]. Among the cases presenting with atrial fibrillation, the case reports by Di Rocco et al. in 2011 remain highly insightful: both patients were of the pediatric age group (14- and 16-year-old males), with a history of acute consumption of large amounts of energy drinks, followed by sensation of palpitations later documented to be atrial fibrillation. Although one patient had a history of baseline stimulant medication use (amphetamine and dextroamphetamine), the effects of acute ingestion of energy drinks in otherwise caffeine-naïve patients are well documented in this case series [20]. Another reported case by Ward et al. (2014) documents an incident of a near-fatal ventricular arrhythmia in a patient with repaired tetralogy of Fallot who was luckily saved by his implanted intracardiac defibrillator (ICD) demonstrating the arrhythmogenic potential of energy drinks in patients with pre-existing cardiac lesions [21]. Other presentations may be less dramatic, such as the previously mentioned cases of QT prolongation. Gray and associates recently revisited concerns of energy drinks' arrhythmogenic potential when they conducted a study that demonstrated 12.5% of their test patient experienced a prolongation of their correction QT interval by >50 ms, further advancing worries of the deleterious effects of energy drinks and their ability to induce life-threatening arrhythmias [22]. Given the current weight of evidence, it is prudent to further delve into the possible explanations on the arrhythmogenic potentials of energy drinks. Caffeine, as has been previously described in Chap. 13, is a well-known arrhythmogenic substance, particularly when taken acutely in large doses [20]. Goldfarb states that caffeine, either due to its direct and indirect effects on the heart or its interaction with other co-ingested substances, is often the common culprit in these drinks causing the witnessed arrhythmias and cardiovascular events. However, the consistent relationship of caffeine with arrhythmogenesis is not entirely perfect. As demonstrated in several large studies, namely, the Women's Health Study by Conen et al. (2010) and the Framingham Heart Study on dietary influences on atrial fibrillation by Shen et al. (2011), chronic consumption of higher amounts of caffeine produced no increased incidence of atrial fibrillation [23, 24]. Therefore, apart from the effects of caffeine, what are the potential arrhythmogenic effects of the other additives found in energy drinks?

Schaffer et al. presented several case reports where interactions between taurine and caffeine were suspected in causing arrhythmias. Prolonged QT interval, atrial fibrillation, and atrial flutter were among the arrhythmias documented among four cases with one episode of postural tachycardia syndrome all of which were documented in patients with varying amounts of energy drink consumption [14]. Thankfully, however, Schaffer states that the arrhythmogenic effects of taurine are poorly demonstrated in humans (as opposed to animal models) and that most of the theoretical worries of acute taurine overdosing is largely unfounded. It is also pointed out that a majority of the evidence present in literature is confined mostly to case reports following a temporal profile of cardiac events in patients following ingestion of energy drinks [14]. Taurine, as a matter of fact, has actually yielded promising results in its function as an anti-arrhythmic agent with promising results in small-scale trials [19, 25]. Ginseng, as previously stated, has an excellent track record of safety as documented in several large trials [18]. However, a case report by Torbey (2011) et al. documents a case of prolonged QT interval with subsequent torsade de pointes in a woman who consumed large amounts of ginseng [26] which may perhaps serve as a warning for the arrhythmogenic potentials of large-dose acute consumption of ginseng and, thus, energy drinks. Other components of energy drinks derive their effect and side effects due to their caffeine contents and have proven to be difficult to analyze in isolation.

17.4 Experimental Induction of Arrhythmias with Energy Drinks

As the case reports presented in the previous section have described, the author notes that most, if not all instances of energy drink induced arrhythmias were established via a temporal profile in the patient's history—that is: documented consumption of increased amounts of energy drinks followed by the occurrence of various types of reported arrhythmias. Such patients were then appropriately treated in emergency departments and hospital settings via standard arrhythmia protocols, leaving little room for confirmation of cause and effect by means of more sophisticated assays, which is understandable, given the urgent circumstance of the arrhythmias presented. This has thus led the author to search further into literature that presents attempts to directly ascribe energy drink consumption to arrhythmias.

An exhaustive investigation of current literature has yielded largely mixed results. The highly detailed meta-analysis by Zuchinali et al. [27] presents an excellent summary of current evidence showing experimental trials that attempted to demonstrate arrhythmias caused primarily by the substance often regarded as the culprit in most energy drinks—caffeine. The meta-analysis summarized a number of animal and human studies that studied the effects of administered, controlled doses of caffeine and analyzed electrophysiologic (EP) end points—particularly the induction of various forms of arrhythmias. The analysis has shown that caffeine, when administered in higher doses or via an intravenous infusion, consistently induces ventricular premature beats (VPBs), increases risk for ventricular fibrillation, and triggers various forms of arrhythmias in animal studies. However, when similar human trials were conducted, involving administration of caffeine (Dose range 175–450 mg) with the goal of achieving EP changes and arrhythmias, the results demonstrated no significant overall electrophysiologic changes [27]. However, a quick look at the studies presented in the paper by Zuchinali, as aptly pointed out in the analysis, may lead one to criticize the quality of human trials. Such trials were noted to be relatively small with varying EP endpoints established [27]. The authors thus emphasize that this particular aspect may benefit from further large-scale studies in the future.

Several other recent studies have been authored that actually seek to challenge the notion presented by Zuchinali (Table 17.3). Firstly, the study by Merrill [28] presented preliminary data that shows significant electrophysiologic (EP) changes demonstrated in young individuals made to consume their preferred energy drink. Analysis for EP and cardiovascular (CV) endpoints were done at 15 min intervals, which demonstrated reduced peripheral blood flow, unstable isoelectric lines on ECG, high amplitude T waves, and deviations of subjects' ST segments [28]. The study by Fletcher et al. [29] conducted among healthy adults recruited from a US air force base likewise demonstrated significant EP and CV results post consumption: there was noted prolongation of QT intervals as well as increases in systolic blood pressure [29].

17.5 Summary

Energy drinks are highly popular beverage formulations that are being consumed at increasing rates. Their components include high amounts of caffeine and other additives that promise enhanced physical and mental performance. Numerous

Table 17.3 Study design and protocols used in arrhythmia induction studies

	Merrill et al. [28]	Fletcher et al. [29]
Population characteristics	Volunteer young subjects with no described comorbidities; Disclosed to be habitual consumers of energy drinks	Healthy young adults recruited from US air force base installation with no comorbidities, no underlying arrhythmias, negative history of drug abuse, and no other over-the-counter drug intake
Sample size	25	18
Mean age	Not specified—described to be college age volunteers	26.7 ± 4 years
Experimental substance used (caffeine concentration)	Subject preferred caffeinated beverage (220 ± 85 mg)	Unlabelled commercially available energy drink (320 mg)
Electrophysiology end points examined (methodology used)	General ECG changes (serial ECG tracings post consumption at 15 min intervals for 60 min)	Primary: Corrected QT interval; Secondary: Uncorrected QT interval, PR interval, QRS duration (serial ECG tracings post consumption taken 1-, 2-, 4-, 6-, and 24-h post consumption)
Other end points examined	Peripheral blood flow	Peripheral systolic blood pressure (pSBP), peripheral diastolic blood pressure (pDBP), central systolic and diastolic blood pressure, augmentation index (AI)
Noted results	Reduction in peripheral blood flow, unstable wandering isoelectric lines on ECG, elevated high amplitude T waves, and deviations of ST segment	Significant change in QT intervals 2 h post consumption, significant change in pSBP and AI 6 h post consumption

case reports and published case series are prevalent in literature that documents the occurrence of fatal and near fatal arrhythmias after consumption of such beverages in large amounts. However, despite such reports, trials carried out on human subjects over the past few years that seek to induce arrhythmias with large amounts of caffeine have largely been negative. Nonetheless, as the author has demonstrated in two relatively new trials, newer evidence has emerged with more updated methodologies that show significant EP changes after consumption of energy drinks. Thus, future research with more stringent methods among larger populations may be required to conclusively and convincingly allay the ever present fear of energy drink triggered arrhythmias. Nonetheless, like with all commercially available substances, the author advises consumption of such beverages in moderation and restraint.

References

1. Kumar G, Park S, Onufrak S. Perceptions about energy drinks are associated with energy drink intake among U.S. youth. Am J Health Promot. 2015;29:238–44.
2. Froiland K, Koszewski W, Hingst J, Kopecky L. Nutritional supplement use among college athletes and their sources of information. Int J Sport Nutr Exerc Metab. 2004;14:104–20.
3. Heckman MA, Sherry K, De Mejia EG. Energy drinks: an assessment of their market size, consumer demographics, ingredient profile, functionality, and regulations in the United States. Compr Rev Food Sci Food Saf. 2010;9:303–17.
4. Goldfarb M, Tellier C, Thanassoulis G. Review of published cases of adverse cardiovascular events after ingestion of energy drinks. Am J Cardiol. 2014;113:168–72.
5. Heneman K, Zidenberg-Cherr S. Nutrition and health info-sheet for health professionals—some facts about energy drinks; Information Pamphlet. Department of Nutrition, University of California; 2007.
6. Clauson KA, Shields KM, McQueen CE, Persad N. Safety issues associated with commercially available energy drinks. J Am Pharm Assoc. 2008;48(3):e55–63.
7. Finnegan D. The health effects of stimulant drinks. Br Nutr Found Nut Bull. 2003;28:147–55.
8. Heck CI, de Mejia EG. Yerba mate tea (Ilex paraguariensis): a comprehensive review on chemistry, health implications, and technological considerations. J Food Sci. 2007;72:138–51.
9. Huxtable RJ. Physiological actions of taurine. Physiol Rev. 1992;72:101–63.
10. Sved DW, Godsey JL, Ledyard SL, Mahoney AP, Stetson PL, Ho S, Myers NR, Resnis P, Renwick AG. Absorption, tissue distribution, metabolism and elimination of taurine given orally to rats. Amino Acids. 2007;32:459–66.
11. Azuma J, Sawamura A, Awata N, Ohta H, Hamaguchi T, Harada H, et al. Therapeutic effect of taurine in congestive heart failure: a double-blind crossover trial. Clin Cardiol. 1985;8:276–82.
12. Takihara K, Azuma J, Awata N, Ohta H, Hamaguchi T, Sawamura A, et al. Beneficial effect of taurine in rabbits with chronic congestive heart failure. Am Heart J. 1986;112:1278–84.
13. Ohta H, Azuma J, Awata N, Hamaguchi T, Tanaka Y, Sawamura A, et al. Mechanism of the protective action of taurine against isoprenaline induced myocardial damage. Cardiovasc Res. 1988;22:407–13.
14. Schaffer SW, Shimada K, Jong CJ, Ito T, Azuma J, Takahashi K. Effect of taurine and potential interactions with caffeineon cardiovascular function. Amino Acids. 2014;46(5):1147–57.
15. Katan MB, Schouten E. Caffeine and arrhythmia. Am J Clin Nutr. 2005;81(3):539–40.
16. de Andrade F, de Albuquerque CA, Maraschin M, da Silva EL. Safety assessment of yerba mate (Ilex paraguariensis) dried extract: results of acute and 90 days subchronic toxicity studies in rats and rabbits. Food Chem Toxicol. 2012;50(2):328–34.
17. Lee N-H, Son CG. Systematic review of randomized controlled trials evaluating the efficacy and safety of ginseng. J Acupunct Meridian Stud. 2011;4:85–9.
18. Eby G, Halcomb WW. Elimination of cardiac arrhythmias using oral taurine with l-arginine with case histories: hypothesis for nitric oxide stabilization of the sinus node. Med Hypotheses. 2006;67:1200–4.
19. Grasser EK, Yepuri G, Dulloo AG, Montani J-P. Cardio- and cerebrovascular responses to the energy drink red bull in young adults: a randomized cross-over study. Eur J Nutr. 2014;53(7):1561–71.
20. Di Rocco JR, During A, Morelli PJ, Heyden M, Biancaniello TA. Atrial fibrillation in healthy adolescents after highly caffeinated beverage consumption: two case reports. J Med Case Rep. 2011;5:18.
21. Ward AE, Lipshultz SE, Fisher SD. Energy drink-induced near-fatal ventricular arrhythmia prevented by an intracardiac defibrillator decades after operative "repair" of tetralogy of Fallot. Am J Cardiol. 2014;114:1124–5.
22. Gray B, Ingles J, Media C, Driscoll T, Semsarian C. Cardiovascular effects of energy drinks in familial long QT syndrome: a randomized cross-over study. Int J Cardiol. 2016;231:148–52.

23. Shen J, Johnson VM, Sullivan LM, Jacques PF, Magnani JW, Lubitz SA, Pandey S, Levy D, Vasan RS, Quatromoni PA, Junyent M, Ordovas JM, Benjamin EJ. Dietary factors and incident atrial fibrillation: the Framingham Heart Study. Am J Clin Nutr. 2011;93:261–6.
24. Conen D, Chiuve SE, Everett BM, Zhang SM, Buring JE, Albert CM. Caffeine consumption and incident atrial fibrillation in women. Am J Clin Nutr. 2010;92:509–14.
25. Gordeev IG, Luchinka EE, Pokrovskaya EM. Taurine effects on the occurrence of cardiac arrhythmias and QT interval dispersion in patients with postinfarction cardiosclerosis and chronic heart failure: a comparative randomised study. Cardiovasc Ther Prevent. 2012;11:63–8.
26. Torbey E, Abi Rafeh N, Khoueiry G, Kowalski M, Bekheit S. Ginseng: a potential cause of long QT. J Electrocardiol. 2011;44:357–8.
27. Zuchinali P, Ribeiro P, Pimentel M, da Rosa P, Zimerman L, Rohde L. Effect of caffeine on ventricular arrhythmia: a systematic review and meta-analysis of experimental and clinical studies. Europace. 2016;18(2):257–66.
28. Merrill GF. Experimentally-induced ventricular arrhythmias. Int J Physiol Pathophysiol Pharmacol. 2017;9(6):202–10.
29. Fletcher E, Lacey C, Aaron M, Kolasa M, Occiano A, Shah S. Randomized controlled trial of high-volume energy drink versus caffeine consumption on ECG and hemodynamic parameters. J Am Heart Assoc. 2017;6(5). pii: e004448.

18 Ajmaline, Flecainide and Propafenone Can Induce Ventricular Fibrillation in Patients with Brugada Syndrome

Natalia Petcaru

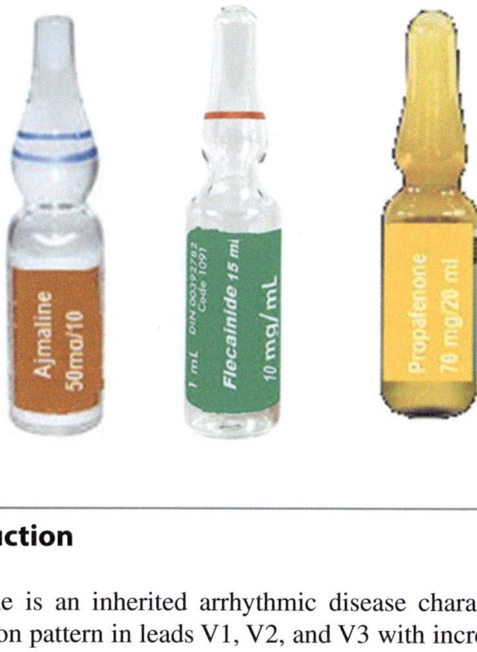

18.1 Introduction

Brugada syndrome is an inherited arrhythmic disease characterized by a cove-shaped ST elevation pattern in leads V1, V2, and V3 with increased risk of sudden cardiac death, without any associated structural heart disease [1]. In patients with suspected Brugada syndrome, intravenous administration of Class IC antiarrhythmic drugs is a well-known method to unmask the syndrome in cases with non-diagnostic type ECG. The most effective among the three drugs is ajmaline, a potent sodium channel blocker having a short half-life [2].

N. Petcaru (✉)
Cardiologie et Maladie Vasculaires, Hopital Avicenne, AP-HP, Bobigny, France
e-mail: natalia-vasilica.pop@aphp.fr

G. Cismaru (ed.), *Arrhythmia Induction in the EP Lab*,
https://doi.org/10.1007/978-3-319-92729-9_18

During injection of ajmaline, life-threatening arrhythmias can occur, like polymorphic ventricular tachycardia or ventricular fibrillation [3]. However, intravenous administration of antiarrhythmic drugs like ajmaline, flecainide, propafenone, procainamide, or disopyramide remains a critical stage in the diagnostic approach of concealed Brugada syndrome.

The presence of Brugada syndrome in pediatric population is low 0.009% compared with adult population 0.14–0.7% [4], but the experience of ventricular fibrillation in a child during drug challenge is always a dramatic event.

18.2 Spontaneous Ventricular Fibrillation After Ajmaline

Ajmaline injection in patients with Brugada syndrome can lead to ventricular tachycardia or ventricular fibrillation, sometimes with intractable episodes. Accordingly, the test should be performed in electrophysiological labs equipped with a defibrillator, a reanimation kit with intubation facilities, mechanical ventilator, and even the possibility to perform a veno-arterial extracorporeal membrane oxygenation placement for severe recurrent episodes of ventricular fibrillation.

In the study of McMillan et al., ajmaline was injected in 95 children with suspicion of Brugada syndrome (mean age = 12 years). In 20% of the patients, a type I ECG was unmasked. No sustained ventricular arrhythmia was observed during ajmaline injection. No patient presented at 3.6 years follow-up arrhythmic events or sudden cardiac death.

On the contrary, in the study of Conte et al., the incidence of ajmaline-induced ventricular arrhythmias was 4.4% in children less than 18 years old, positive for ajmaline challenge test. In total the authors injected ajmaline in 503 patients and 9 of them presented life-threatening ventricular arrhythmias, defined as refractory to the first external defibrillation. Six of the patients had ventricular fibrillation and three unstable polymorphic ventricular tachycardia. One of the patients had prolonged intractable VF with multiple unsuccessful defibrillations >10 and necessitated ECMO to restore sinus rhythm. She presented cerebral edema and neurologic sequelae and acute respiratory distress syndrome [5]. The mean dose injected in their study was 0.6 ± 0.2 mg/kg (total dose 34 ± 20 mg).

Brugada et al. report one case of ventricular fibrillation during injection for drug challenge in BS on 45 patients. Thirty-four had BS, and 11 were family members with positive SCN5A and received 1 mg/kg ajmaline (Fig. 18.1), 10 mg/kg procainamide, or 2 mg/kg flecainide. One of the patients that received ajmaline presented spontaneous ventricular fibrillation.

Rolf et al. also reported ventricular arrhythmias in 1.3% of 158 patients that received ajmaline. The explanation could be the continuation of infusion even after occurrence of clear signs of Brugada syndrome type 1 on the surface ECG. Therefore, termination criteria were implemented for the challenge test for all the three abovementioned medications (Table 18.2).

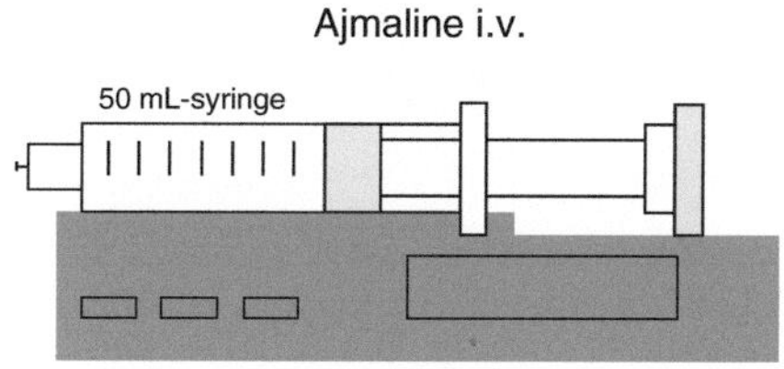

Fig. 18.1 Ajmaline dose during challenge test

Bermudez et al. reported the case of a 13-year-old patient with Brugada syndrome and sudden cardiac arrest by ventricular fibrillation. He presented severe postanoxic encephalopathy. Two weeks after the event an electrophysiological study was performed with administration of 1 mg/kg ajmaline. The patient developed runs of nonsustained VT with a pleomorphic morphology and sustained monomorphic ventricular tachycardia.

18.2.1 Protocol of Administration

Arnalsteen-Dassonvalle et al. compared two protocols of ajmaline 1 mg/kg administration: a rapid protocol of infusion at 1 mg/s and a slow protocol over 10 min. They found no significant difference between the rapid and the slow protocols of infusion. No sustained ventricular arrhythmia was observed in the two groups. PVCs were observed in five of 32 patients from the slow challenge group and four of 32 patients from the rapid challenge group (Table 18.1).

We believe that the slow protocol is safer, because ajmaline might be interrupted if ventricular fibrillation occurs, and only a part of the total dose of 1 mg/kg would be injected.

Therasse et al. studied 672 patients, including 175 patients (26%) that received flecainide. Ventricular tachycardia and fibrillation occurred in 10 of the 497 patients that received ajmaline and none of the 175 patients that received flecainide [6].

Dobbels et al. analyzed all the studies published between January 2000 and August 2015 which reported ventricular arrhythmias after ajmaline. They found three predictors for malignant ventricular arrhythmia occurrence: younger age, conduction disturbance at the baseline ECG, and mutation in the SCN5A gene [7].

Gandjbakhch et al. reported the results of ajmaline challenge test in a large family with SCN5A mutation. A 34-year-old patient (Table 18.3) presented during ajmaline 1 mg/kg at 1 mg/s a polymorphic ventricular tachycardia associated with syncope. Two relatives of the patient received the same dose of ajmaline during the text and presented polymorphic ventricular tachycardia with syncope [8]. It is unlikely that this response be caused by the rapid infusion protocol, as a randomized study on 336 patients found that a rapid infusion protocol is not associated with a higher risk of ventricular arrhythmias [9].

Table 18.1 Ajmaline test for Brugada syndrome

1 mg/kg ajmaline	During 10 min
ECG	At 1 min
ECG	At 2 min
ECG	At 3 min
ECG	At 4 min
ECG	At 5 min
ECG	At 6 min
ECG	At 7 min
ECG	At 8 min
ECG	At 9 min
ECG	At 10 min
ECG	At 15 min
ECG	At 20 min
ECG	At 25 min
ECG	At 30 min

Table 18.2 Termination criteria for ajmaline test

1. A typical type-1 Brugada pattern (V1–V3)
2. Ventricular arrhythmias (VEs or VT]
3. QRS duration prolongs >30% from baseline
4. Type 2 or 3 AV block
5. Maximum dose given

18.2.2 Termination Criteria

See Table 18.2.

18.2.3 Reversal of Ajmaline with Isoprenaline

Isoproterenol stimulating β-receptors augments Ica and reduces ST segment elevation in leads V1–V3 in patients with BS. Isoproterenol was demonstrated to be effective in VF suppression in a 36-year-old male with Brugada syndrome and determined disappearance of the short-coupled PVCs, which can be triggers of ventricular fibrillation [10]. The effectiveness of isoproterenol was also confirmed by Watanabe et al. in a case series of patients with Brugada syndrome in which ventricular severe arrhythmias were successfully abolished with infusion of isoproterenol in six patients. In case of PVCs, ventricular tachycardia or ventricular fibrillation during the ajmaline test, isoprenaline should be administered. As an established protocol, when modifications of the ST segment occur, isoprenaline is injected at the end of the electrophysiological study to monitor modifications of the ST. The dose of isoprenaline is 2 μg/min over 10–30 min with continuous ECG monitoring.

In the study of Arnalsteen-Dassonvalle et al., patients received a higher dose of isoproterenol than usually described in electrical storms: 0.06 mg/min compared to 0.1–1 μg/min [11, 12].

18.2.4 Reversal of Ajmaline with Sodium Bicarbonate

Sodium bicarbonate is a good treatment for sodium channel blockers overdose and is indications for cardiac arrest, increased widening of QRS complex, and cardiogenic shock refractory to fluid therapy. After sodium bicarbonate the QRS duration decreases, and probably the QRS will normalize. The efficacy of the sodium bicarbonate is demonstrated by animal studies and human reports both in adult and in pediatric population. The effectiveness is likely mediated by the alkalemia induced by the bicarbonate and release of sodium ions to fast-acting sodium channels inside the affected myocardium.

Sodium bicarbonate is injected in doses of 150 mEq/L, 8.4% solution (Fig. 18.2).

18.2.5 ECMO for Ventricular Fibrillation After Ajmaline

Poli et al. reported the case of a 32-year-old woman that presented polymorphic ventricular tachycardia and ventricular fibrillation without return to normal rhythm after 14 DC shocks. The patients was transferred to another hospital 50 km away and treated with extracorporeal membrane oxygenation. After 30 min of ECMO transvenous ventricular pacing was effective, and the patient further recovered normal sinus rhythm [13]. Survival of patients with cardiac arrest and prolonged time to return to spontaneous circulation shows to be rare. ECMO is an

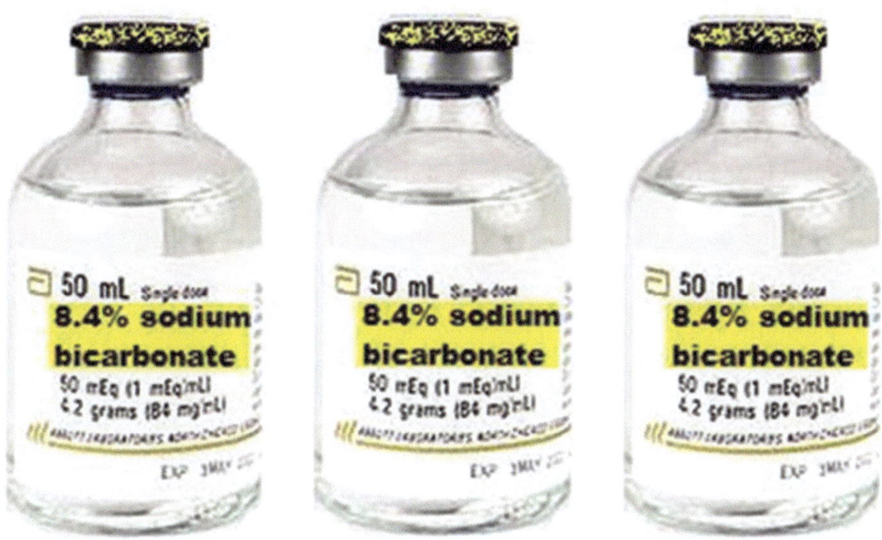

Fig. 18.2 Sodium bicarbonate dose during ventricular arrhythmias induced by ajmaline testing (3 × 50 mg intravenously)

Table 18.3 How to prepare the ajmaline infusion for the challenge test

Understanding the numbers:

You are ordered to give 1 mg/kg of Ajmaline over 5 min to your 58-year-old, 70 kg patient with suspicion of Brugada Syndrome

1. The aminophylline vial has 50 mg/10 mL
2. You should take 2 vials of Ajamaline in the 50 mL syringe with 30 mL of serum (0.9% NaCl)
3. This gives a concentration of 100 mg/50 mL meaning 2 mg/mL
4. 1 mg x 70 kg=70 mg should be injected, equivalent of 35 mL
5. We will start the perfusion rate: 35x12= 420 mL/h

option in case of refractory cardiogenic shock or cardiac arrest in patients that have a reversible underlying pathology. Acute toxic exposure is an indication for ECMO because restoring circulation permits the intrinsic drug metabolism and elimination from the circulation. The problem with ECMO is the limited number of centers that can implant ECMO and manage patients having this support. ECMO can improve outcomes of survival and morbidity at 1 year follow-up associated with advanced life support, when the technique is indicated. Sodium channel blockers challenged should be performed in electrophysiological labs that are equipped with CPR facilities available and ECMO in the same hospital or nearby area. In the current medical literature, there are at least four case reports with ECMO following severe flecainide intoxication.

18.3 Spontaneous Ventricular Fibrillation After Flecainide

Flecainide test is widely used in patients with Brugada syndrome, especially in countries where ajmaline is not available. The ability to identify patients with concealed form of Brugada syndrome is clearly established [14].

Gasparini et al. injected flecainide in 22 patients with BS, eight of them with SCN5A. The dose infused was 2 mg/kg over 10 min. Three patients developed sustained ventricular tachycardia with duration of 7–10 min which ended spontaneously and recurrent ventricular fibrillation in three patients. The reproducibility of the test was 100%. Ventricular arrhythmias occurred especially in patients with SCN5A mutation 3/7 than without mutation 1/15.

Therasse et al. studied 672 patients, including 175 patients (26%) that received flecainide. Ventricular tachycardia and fibrillation occurred in 10 of the 497 patients that received ajmaline and none of the 175 patients that received flecainide.

In the study of Brugada et al., 2 mg/kg flecainide was infused over 5 min. Spontaneous ventricular fibrillation occurred in 1 of 53 patients and 5 patients presented frequent PVCs during infusion.

18.3.1 Challenge Test with Oral Flecainide

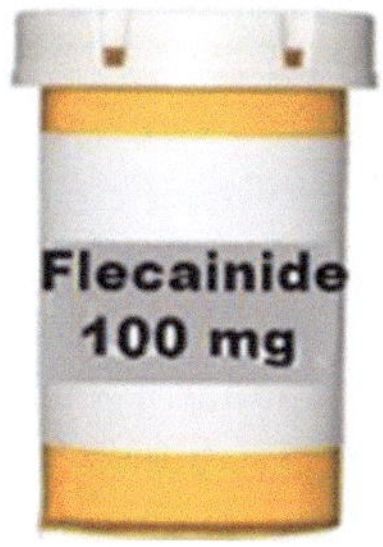

The challenge test can be also made with oral flecainide (Tables 18.4 and 18.5). The dose is 400 mg in single administration. Bioavailability in oral administration is around 70% (60–86%), and a higher bioavailability can be obtained with a higher dose.

In the study of Prasad et al. out of 29 patients 25 received oral flecainide 400 mg, the test was positive in 7 patients, and none presented spontaneous ventricular

Table 18.4 Flecainide test with tablets

Two tablets of 200 mg	400 mg
ECG	At 15 min
ECG	At 30 min
ECG	At 60 min
ECG	At 90 min
ECG	At 2 h
ECG	At 3 h
ECG	At 4 h
ECG	At 5 h
ECG	At 6 h

Table 18.5 How to prepare the flecainide infusion for a challenge test

Understanding the numbers:

You are ordered to give 2 mg/kg of Flecainide over 10 min to your 58-year-old, 70 kg patient with suspicion of Brugada Syndrome

1. The Flecainide vial has 15 mL and 10 mg/1 mL equivalent of 150 mg
2. You should take 1 vial of Flecainide in the 50 mL syringe with 35 mL of serum (0.9% NaCl)
3. This gives a concentration of 150 mg/50 mL meaning 3 mg/mL.
4. 2 mgx70 kg=140 mg should be injected, equivalent of 46.6 mL
5. We will start the perfusion rate: 46.6x6= 280 mL/h

arrhythmia. The authors observed a maximum time to positivity of 3 h and a maximum time to normalization of ECG of 6 h. Therefore, they recommend a maximum observation time of 6 h.

18.4 Spontaneous Ventricular Fibrillation After Propafenone

Not only ajmaline, flecainide and procainamide can unmask Brugada syndrome but also propafenone which is also a class IC antiarrhythmic drug (Fig. 18.3, Table 18.6). It blocks the sodium channels and also it blocks the β-receptors. Both sodium-blocking agents and beta-blockers can unmask Brugada syndrome.

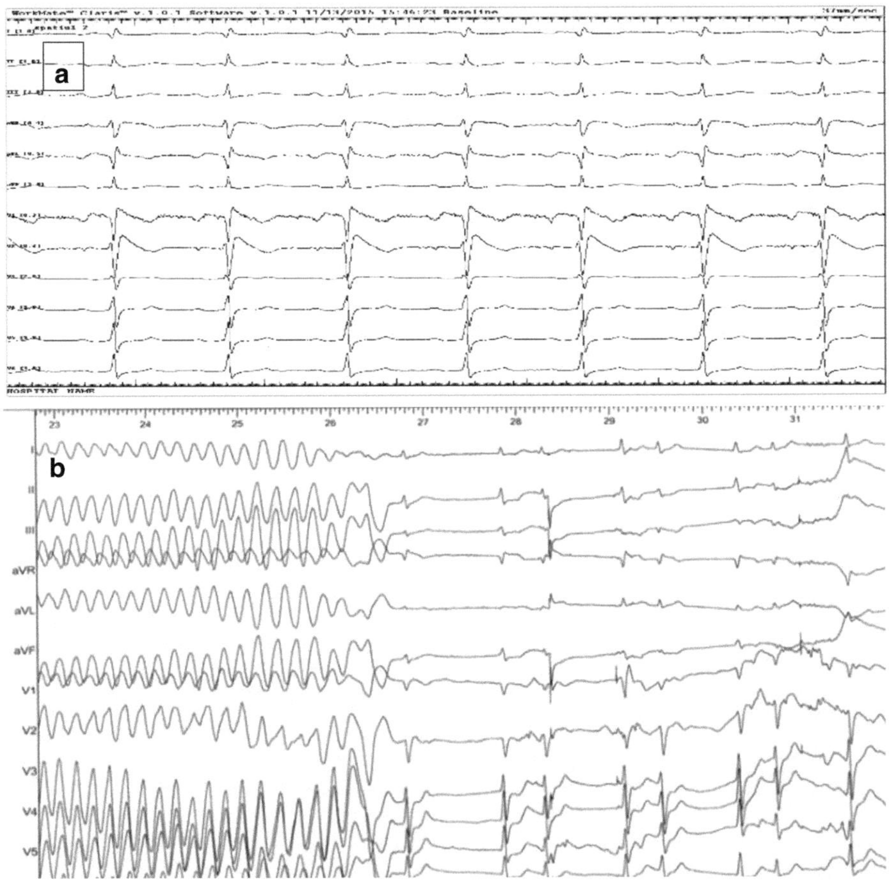

Fig. 18.3 Propafenone challenge test was performed with suspicion of Brugada syndrome. Four hundred milligrams was orally administered, and the patient developed Brugada pattern (**a**). One hour after propafenone the patient presented non-sustained ventricular fibrillation (**b**)

Table 18.6 How to prepare the propafenone infusion for a challenge test

Understanding the numbers:

You are ordered to give 2 mg/kg of Propafenone over 10 min to your 58-year-old, 70 kg patient with suspicion of Brugada Syndrome

1. The Propafenone vial has 70 mg and 20 mL
2. You should take 2 vials of Flecainide in the 50 mL syringe with 10 mL of serum (0.9% NaCl)
3. This gives a concentration of 140 mg/50 mL meaning 2,8 mg/mL.
4. 2 mgx70 kg=140 mg should be injected, equivalent of 50 mL
5. We will start the perfusion rate at 50 mL/h

Aksay et al. reported the case of a 43-year-old man who received 600 mg of propafenone for conversion of atrial fibrillation. ST segment elevation occurred in leads V1–V3, confirming the Brugada syndrome. Modifications disappeared from the ECG after 6 h.

Matana et al. presented the first case of concealed Brugada syndrome that was unmasked by propafenone [15].

In 1997, there were described the first cases of Brugada syndrome with ventricular fibrillation spontaneous induction after propafenone administration [16]. Kose et al. reported the case of a patient that presented ventricular fibrillation during propafenone challenge test. A 46-year-old patient had repeated episodes of syncope (3 episodes). He had no familial history of syncope or sudden cardiac death. During provocation propafenone test, 2 mg/kg were injected intravenously during 10 min. The type 2 pattern present in the beginning of the test transformed in type 1 at 7 min interval after infusion, and after 2 more min, the patient developed ventricular fibrillation. The patient returned to sinus rhythm after an electrical shock of 200 J. The patient was implanted with an internal defibrillator [17].

Rodriguez-Manero et al. from the Brugada group reported 611 patients with Brugada syndrome of whom 35 presented atrial fibrillation as the first clinical manifestation of the syndrome. Eleven patients developed ECG pattern after class IC antiarrhythmic drug for atrial fibrillation, one patient presented sudden cardiac death after propafenone administration, and one presented ventricular fibrillation after flecainide [18].

References

1. Brugada P, Brugada J. Right bundle branch block, persistent ST segment elevation and sudden cardiac death: a distinct clinical and electrocardiographic syndrome. A multicenter report. J Am Coll Cardiol. 1992;20:1391–6.
2. Antzelevitch C, Brugada P, Borggrefe M, Brugada P, Brugada R, Corrado D, et al. Brugada syndrome: report of the second consensus conference. Circulation. 2005;111:659–70.
3. Wellens HJ, Bar FW, Vanagt EJ. Death after ajmaline administration. Am J Cardiol. 1980;45:905.

4. Wolpert C, Echternach C, Veltmann C, Antzelevitch C, Thomas GP, Spehl S, et al. Intravenous drug challenge using flecainide and ajmaline in patients with Brugada syndrome. Heart Rhythm. 2005;2:254–60.
5. Conte G, Sieira J, Sarkozy A, et al. Life-threatening ventricular arrhythmias during ajmaline challenge in patients with Brugada syndrome: incidence, clinical features, and prognosis. Heart Rhythm. 2013;10:1869–74.
6. Therasse D, Sacher F, Petit B. Sodium-channel blocker challenge in the familial screening of Brugada syndrome: safety and predictors of positivity. Heart Rhythm. 2017;14:1442–8.
7. Dobbels B, De Cleen D, Ector J. Ventricular arrhythmia during ajmaline challenge for the Brugada syndrome. Europace. 2016;18:1501–6.
8. Gandjbakhch E, Fressart V, Duthoit G. Malignant response to ajmaline challenge in SCN5A mutation carriers: experience from a large familial study. Int J Cardiol. 2014;172:256–8.
9. Arnalsteen-Dassonvalle E, Hermida JS, Kubala M, Six I, Quenum S, Leborgne L, et al. Ajmaline challenge for the diagnosis of Brugada syndrome: which protocol? Arch Cardiovasc Dis. 2010;103:570–8.
10. Yan GX, Antzelevitch C. Cellular basis for the Brugada syndrome and other mechanisms of arrhythmogenesis associated with ST-segment elevation. Circulation. 1999;100:1660–6.
11. Jongman JK, Jepkes-Bruin N, Ramdat Misier AR, Beukema WP, Delnoy PP, Oude Lutttikhuis H, et al. Electrical storms in Brugada syndrome successfully treated with isoproterenol infusion and quinidine orally. Neth Heart J. 2007;15:151–5.
12. Maury P, Couderc P, Delay M, Boveda S, Brugada J. Electrical storm in Brugada syndrome successfully treated using isoprenaline. Europace. 2004;6:130–3.
13. Poli S, Toniolo M, Maiani M, Zanuttini D, Rebellato L, Vendramin I, et al. Management of untreatable ventricular arrhythmias during pharmacologic challenges with sodium channel blockers for suspected Brugada syndrome. Europace. 2018;20:234–42.
14. Priori SG, Wilde AA, Horie M, Cho Y, Behr ER, Berul C, et al. Executive summary: HRS/EHRA/APHRS expert consensus statement on the diagnosis and management of patients with inherited primary arrhythmia syndromes. Europace. 2013;15:1389–406.
15. Matana A, Goldner V, Stanic K. Unmasking effect of propafenone on the concealed form of the Brugada phenomenon. Pacing Clin Electrophysiol. 2000;23:416–8.
16. Kasanuki H, Ohnishi S, Ohtuka M, Matsuda N, Nirei T, Isogai R, et al. Idiopathic ventricular fibrillation induced with vagal activity in patients without obvious heart disease. Circulation. 1997;95:2277–85.
17. Kose N, Akın F. Ventricular fibrillation developing during the provocation test with propafenone in a patient with Brugada syndrome. Int J Cardiol. 2013;163:S81–S211.
18. Rodríguez-Mañero M, Namdar M, Sarkozy A, Casado-Arroyo R, Ricciardi D, de Asmundis C, Chierchia GB, Wauters K, Rao JY, Bayrak F, Van Malderen S. Brugada prevalence, clinical characteristics and management of atrial fibrillation in patients with Brugada syndrome. Am J Cardiol. 2013;111:362–7.

Arrhythmia Induction in Children

19

Nikola Krmek

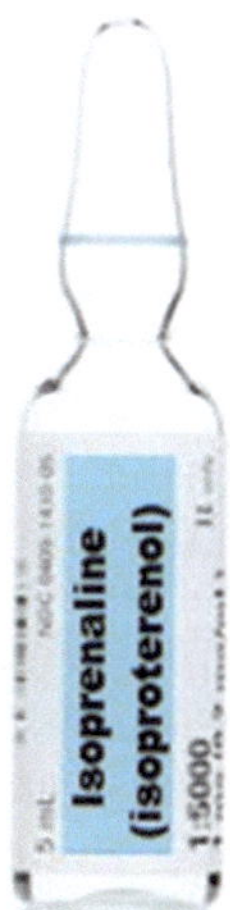

19.1 Introduction

In this chapter we will demonstrate that medication administration in children for arrhythmia induction has specific characteristics. Basic principles such as pharmacology and mechanism of action are mostly the same as in adults, and the biggest difference is in dosing. It goes without saying that there are not many studies performed in children as in adults and many of dosing schemes are based on the particular experience of different electrophysiology centers.

N. Krmek (✉)
Department of Pediatrics, University Clinical Hospital Center "Sestre Milosrdnice", Zagreb, Croatia

G. Cismaru (ed.), *Arrhythmia Induction in the EP Lab*,
https://doi.org/10.1007/978-3-319-92729-9_19

The current practice for selecting the pediatric dose for the drugs used in electrophysiology does not guarantee a safe and effective dose for children. Reducing the dose used in adults by formulas using age, body weight, or any other demographic variable without any prior evidence of how these factors contribute to the drug's plasma concentration may lead to unreliable pediatric dose estimates. However, this practice is universally used in electrophysiology for medication that induces arrhythmias in children as in adults. Although it is relatively easy and simple to use empirical doses for these drugs, growth and development of the child can result in unpredictable dose-effect pharmacokinetic changes. Recent advances in qualitative and quantitative pharmacokinetic analysis will allow electrophysiologists to determine the most appropriate dose for the child that needs ablation. Until then, we should use the doses published by pediatric electrophysiology teams in various articles dealing with pediatric catheter ablation. These doses will be listed in the following paragraphs.

19.2 Isoproterenol for Arrhythmia Induction in Children

There are differences from center to center in isoprenaline administration during electrophysiology studies. Some authors use bolus administration and others continuous infusion. Mixed approach is also possible with bolus injection and then infusion. Studies on dosing of isoprenaline are mostly performed on children with AV block. Infused dose ranges from 0.03 to 0.5 μg/kg/min with titration to effect (20–50% rise in heart rate or more than 130 bpm). Maximum dose in some publication is 1 μg/min although in *larger* children, adult dosing scheme could be applied [1–4]. Bolus dose is usually 0.1 μg/kg.

As the use of isoprenaline is for short duration during an electrophysiology study, major side effects are very rare. However, one has to consider the longer administration, in particular cases when side effects are more likely to occur (from prescribing information of the product: intravenous infusions of isoproterenol in refractory asthmatic children at rates of 0.05–2.7 μg/kg/min have caused clinical deterioration, myocardial necrosis, congestive heart failure, and death). The risks of cardiac toxicity appear to be increased by some circumstances: acidosis, hypoxemia, coadministration of corticosteroids, and coadministration of methylxanthines (theophylline, theobromine) or aminophylline [5].

19.2.1 Doses of Isoprenaline in Children

Liberman et al. infused isoproterenol in 75 patients with narrow complex tachycardia and successful ablation of an accessory pathway. The mean age was 13 years with range between 2.8 and 24 years. After catheter ablation, isoproterenol was infused in dose of 0.01–0.03 μg/kg/min with an increase of 20% of the basal heart rate (Fig. 19.1). During infusion, programmed stimulation was repeated, and three patients presented inducible arrhythmia under isoprenaline: one with recurrent

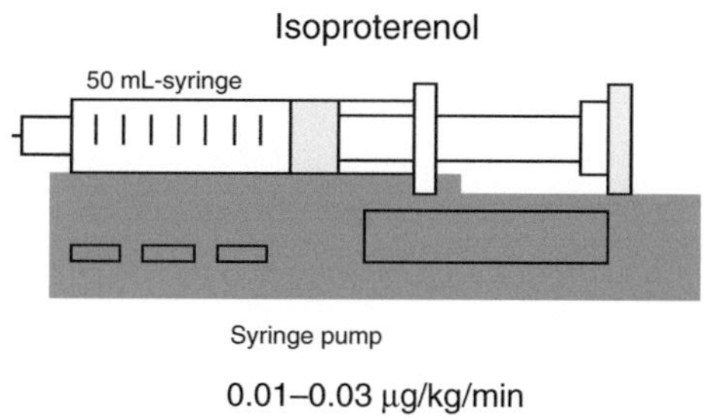

Fig. 19.1 Isoproterenol dose in children (Liberman's study)

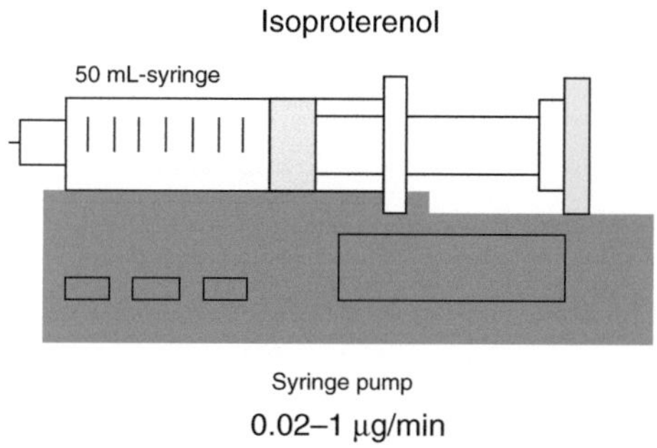

Fig. 19.2 Isoproterenol dose in children (Brembilla-Perrot's study)

accessory pathway and two with AVNRT in which catheter ablation of the slow pathway was performed [5].

Brembilla-Perrot et al. studied 173 children with palpitations evaluated by transesophageal electrophysiological study. In 149 children, supraventricular tachycardia was induced during EP study, and in 124 children arrhythmia was non-inducible. When arrhythmia was non-inducible during transesophageal study, isoproterenol was infused 0.02–1 μg/min to increase the heart rate to at least 130 bpm, and the stimulation protocol was repeated (Fig. 19.2). In the 124 patients with negative electrophysiological study, isoproterenol induced just a sinus tachycardia with a rate of 140–160 bpm that reproduced the symptoms of the child [6].

Mah et al. studied 41 patients with intermittent preexcitation from a database of 328 patients with preexcitation and compared patients having intermittent preexcitation with patients having persistent preexcitation. In 17 patients with intermittent preexcitation and 41 patients with persistent preexcitation, isoproterenol was infused in dose of 0.1 μg/kg bolus over 1 min, followed by a 0.01–0.02 μg/kg/min infusion (Fig. 19.3). Of the 17 patients with intermittent preexcitation, after isoproterenol infusion 3 remained non-preexcited. Electrophysiological testing under isoproterenol added 12 patients from the persistent preexcitation group to the "high-risk group" and 3 patients from the intermittent preexcitation group to the "high-risk" group [7].

Brembilla-Perrot et al. studied 63 patients with antidromic tachycardia. Compared to patients without antidromic tachycardia, in the studied group, the malignant form, left lateral accessory pathway and 1:1 conduction through the accessory pathway, was more frequent, and septal location was less frequent. Atrial fibrillation

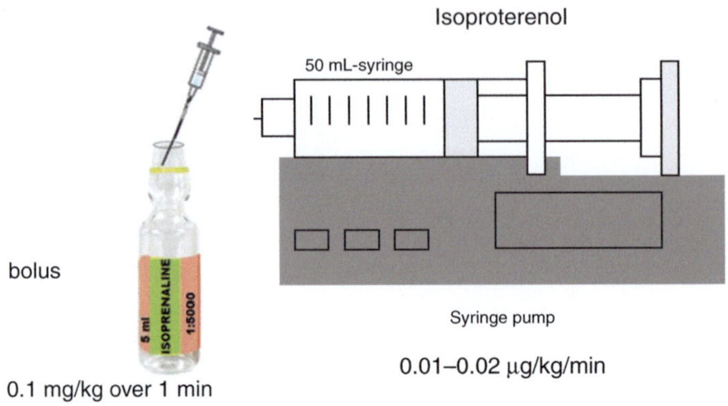

Fig. 19.3 Isoproterenol dose in children: bolus plus infusion (Mah's study)

Fig. 19.4 Isoproterenol dose in children (Brembilla-Perrot's study)

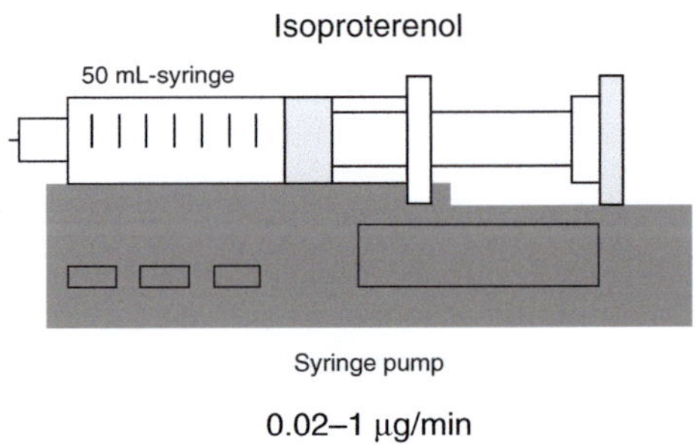

induction was more frequent in the antidromic tachycardia group. Programmed atrial stimulation was performed without and with isoproterenol. The dose used was 0.02–1 µg/min to increase the heart rate to at least 130 bpm. Accessory pathway was considered malignant if 1:1 conduction to the ventricles was >300 bpm after isoproterenol infusion. Antidromic tachycardia was induced in the basal state in 31 patients and only after isoproterenol in 32 patients [8] (Fig. 19.4).

Mills et al. verified the theory that in children ORT has a faster rate than AVNRT. They measured the tachycardia cycle length in 835 patients: 539 with AVRT and 296 with AVNRT. They found that in basal state tachycardia cycle length is shorter in the case of AVRT (329 ms) than in the case of AVNRT (340 ms). After isoproterenol infusion, there was no difference between the two groups: 290 ms for AVRT and 297 ms for AVNRT. Only after controlling for age, authors found no difference between the two groups for the tachycardia cycle length in the basal state without isoproterenol. The dose of isoproterenol was 1 µg/min for patients >10 years of age, 0.5 µg/min for patients 4–10 years of age, and 0.25 µg/min (0.05 µg/kg/min) for patients less than 4 years of age (Figs. 19.5, 19.6, and 19.7). Of the total number of 539 children with SVT, 239 (29%) needed isoproterenol for arrhythmia induction. Isoproterenol was needed to induce AVNRT in 148/296 patients (50%) and AVRT in 91/539 (17%) [9].

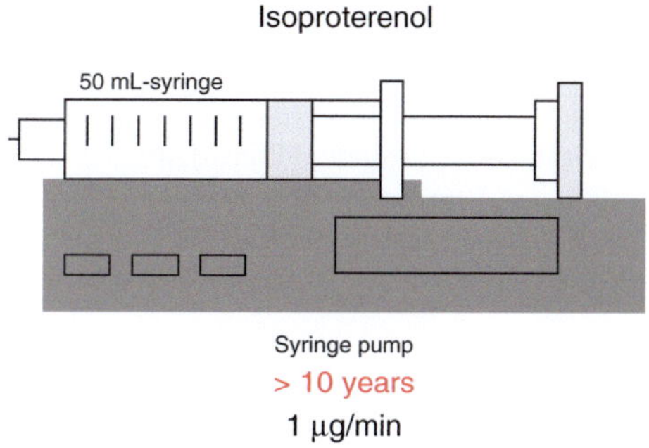

Fig. 19.5 Isoproterenol dose in children >10 years (Mills's study)

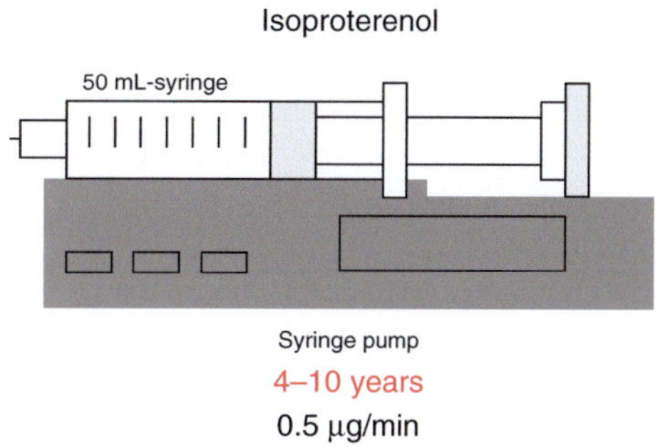

Fig. 19.6 Isoproterenol dose in children 4–10 years (Mills's study)

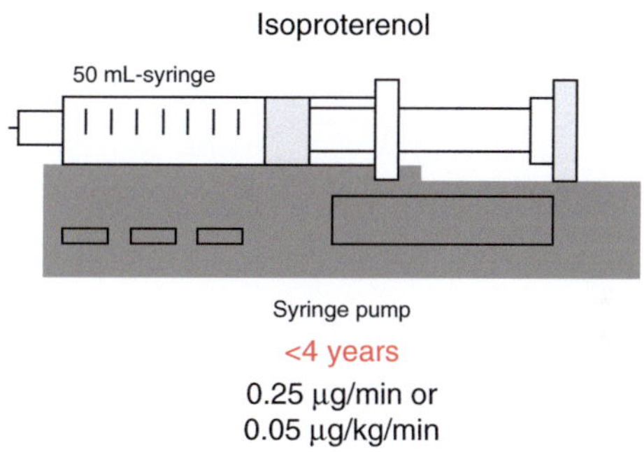

Fig. 19.7 Isoproterenol dose in children <4 years (Mills's study)

In the study of Emmel et al., the authors performed catheter ablation in three patients with ectopic junctional tachycardia using the LocaLisa mapping system. The patients had 7, 9, and 12 years, and tachycardia was inducible after isoproterenol infusion. The dose used by the authors was 0.03 μg/kg/min (Fig. 19.8). Mapping was performed during tachycardia and ablation during sinus rhythm. LocaLisa system permitted to tag the best activation time during tachycardia near the level of the His bundle. The ablation was successful in all three patients without any complications of the His conduction system [10].

Brembilla-Perrot and her group from Nancy (France) performed an electrophysiological study in 140 patients (mean age 15 years) with paroxysmal supraventricular tachycardia. When arrhythmia was non-inducible by programmed atrial and ventricular stimulation, isoproterenol was infused in doses of 0.02–1 μg/min (Fig. 19.9). Eighty-three patients had inducible AVNRT, 1 patient had atrial tachycardia, 33 had ORT, and 4 patients had idiopathic left or right ventricular tachycardia [11].

Drago et al. evaluated 26 patients with AVNRT or accessory pathway treated by cryoablation. A high percentage of patients needed isoproterenol for arrhythmia induction: 12 of 14 patients with AVNRT (86%) and 11 of 12 patients with accessory pathway (92%). The dose infused was 0.01–0.04 μg/kg/min (Fig. 19.10). Of the 26 patients treated with cryoablation, only 2 presented arrhythmia recurrences. The success rate was 92% [12].

Ko et al. studied 67 children and infants with a mean age of 8.3 years with palpitations but without electrocardiographic documentation of the clinical arrhythmia. Forty-seven (70%) of the children had inducible tachycardia, and in ten patients, isoproterenol was infused for arrhythmia induction. The dose of isoproterenol was 0.05–0.1 μg/kg/min (Fig. 19.11). The type of tachycardia induced was AVNRT in 16 patients, ORT in 25 patients, and left ventricular idiopathic tachycardia in 6 patients [13].

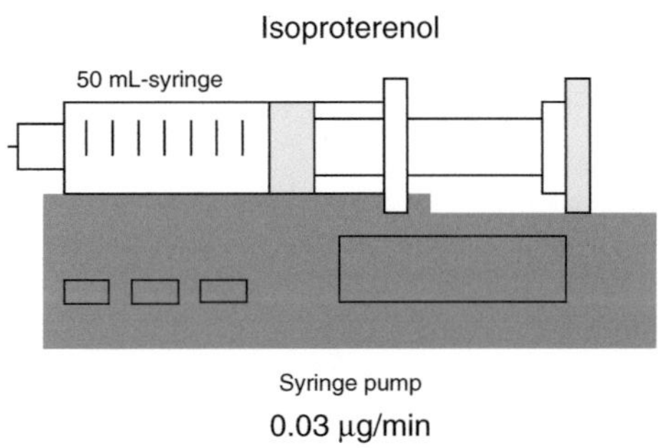

Fig. 19.8 Isoproterenol dose in children (Emmel's study)

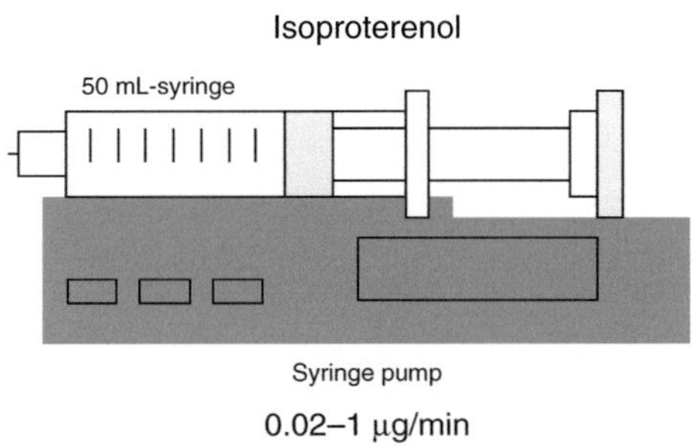

Fig. 19.9 Isoproterenol dose in children (Brembilla-Perrot's study)

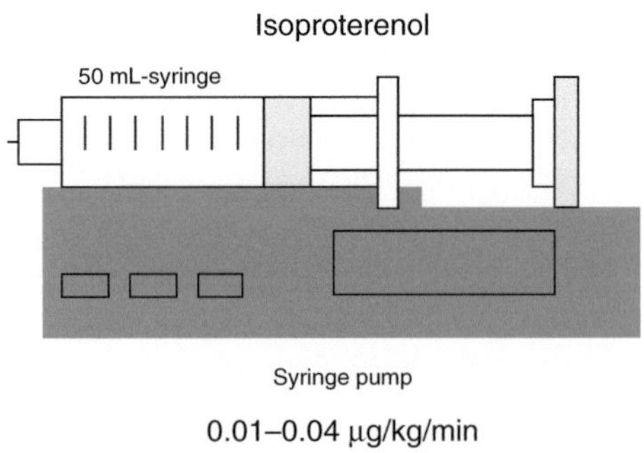

Fig. 19.10 Isoproterenol dose in children (Drago's study)

Law et al. studied the incidence and inducibility of intra-atrial reentrant tachycardia in children before Fontan operation. There were 44 patients with a mean age of 1.7 years and with a mean weight of 11 kg. Intra-atrial reentry was induced by atrial stimulation either with atrial burst or with ramp. If arrhythmia was uninducible, isoproterenol was infused at a rate of 0.01–0.1 μg/kg/min to increase the heart rate with 20% of the basal rate or to more than 150 bpm (Fig. 19.12). Programmed atrial stimulation and burst were repeated [14].

Rhodes et al. performed 203 transesophageal electrophysiological studies to test different treatment options for infants and children with supraventricular tachycardias. The predictive study of a negative transesophageal study was 89% and further increased to 96% after infusion of isoproterenol. The dose was 0.01–0.1 μg/kg/min (Fig. 19.13).

Shih et al. studied five patients with permanent junctional reciprocating tachycardia, and isoproterenol 25–50 μg/kg/min was infused to facilitate atrioventricular

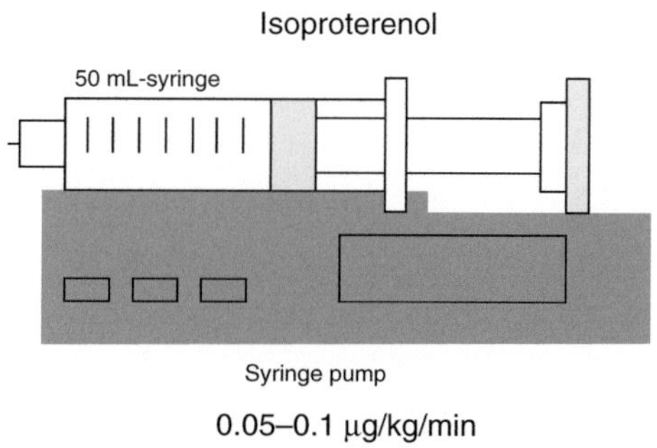

Fig. 19.11 Isoproterenol dose in children (Ko's study)

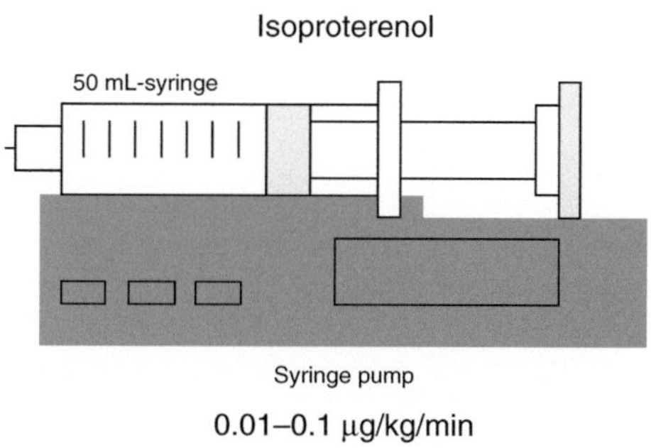

Fig. 19.12 Isoproterenol dose in children (Law's study)

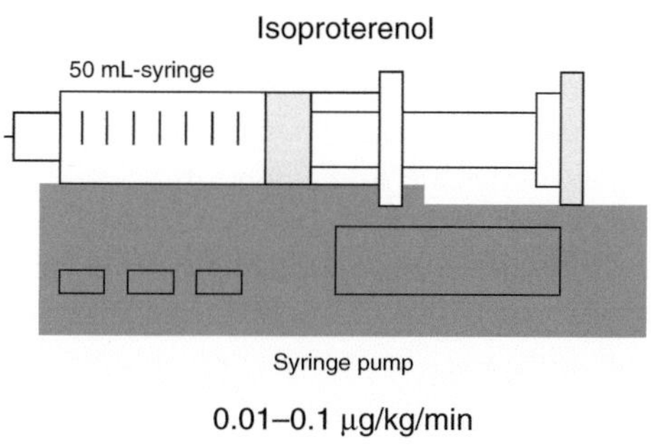

Fig. 19.13 Isoproterenol dose in children (Rhodes's study)

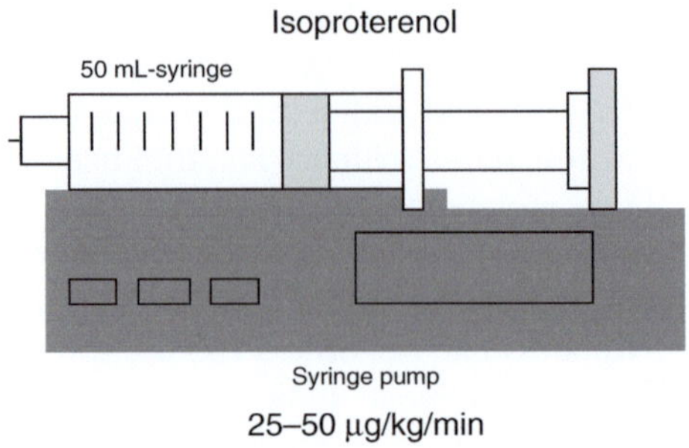

Fig. 19.14 Isoproterenol dose in children (Shih's study)

reentrant tachycardia. Three of the patients had two accessory pathways. Catheter ablation was effective, and the patients remained asymptomatic at 7–29-month follow-up. Four to 10 weeks after the ablation, a new electrophysiological study was performed for non-inducibility. Four patients also received isoprenaline 25–50 µg/kg/min [15] (Fig. 19.14).

19.3 Atropine for Arrhythmia Induction in Children

Atropine is used less frequently in children in the EP lab. There are no prospective studies of effectiveness or dosing in children in electrophysiology. The usual dose is 0.02 mg/kg (ranging from 0.01 to 0.05 mg/kg). The minimum dose is 0.1 mg, and the maximum single dose is 0.5 mg [16–20]. Doses <0.1 mg should be avoided because of the risk of paradoxical bradycardia. It has been used in combined autonomic blockade and unrevealing of slow pathway function in children without major side effects.

In the study of Marcurs et al., atropine was injected in doses of 0.04 mg/kg and propranolol 0.2 mg/kg for total autonomic blockade. Intrinsic sinus node function was evaluated after autonomic blockade (Fig. 19.15). In normal children with a mean age of 9 years, the intrinsic heart rate was 128 bpm, the

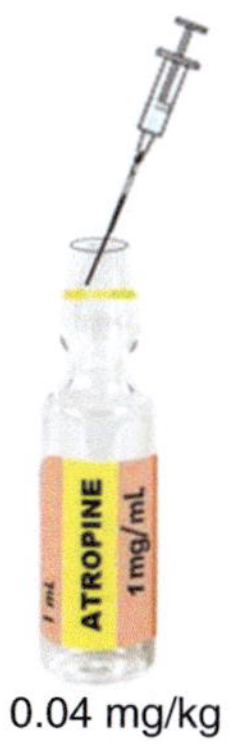

Fig. 19.15 Atropine dose in children (Marcurs's study)

Fig. 19.16 Atropine dose in children (Sileikiene's study)

corrected sinus node recovery time was 135 ms, and the intrinsic sinoatrial conduction time was 86 ms [21].

In the study of Sileikiene et al., anterograde conduction was evaluated before and after catheter ablation of slow pathway in 30 patients using transesophageal electrophysiological study. If AVNRT remained uninducible, atropine was injected in doses of 0.02 mg/kg and the EP study repeated (Fig. 19.16). After catheter ablation, the slow pathway was absent in 17 patients. At 2–8-year follow-up after the initial ablation, transesophageal electrophysiological study was performed, and the slow pathway was present in 28 patients after adenosine administration [22].

Sledz et al. studied the PR/RR ratio in 60 patients, mean age 14 years, with 58% girls. The PR/RR ratio is a simple tool for the presence of slow pathway and a surrogate endpoint for AVNRT ablation. Children received isoproterenol bolus in dose of 0.01–0.03 mg to increase the heart rate with more than 30% of the basal heart rate or more than 100 bpm (Fig. 19.17). If no arrhythmia was induced, the child received 1 mg of bolus atropine. The PR/RR ratio before ablation was 1.14 in the AVNRT group and 0.80 in the control group. In the control group, there were no

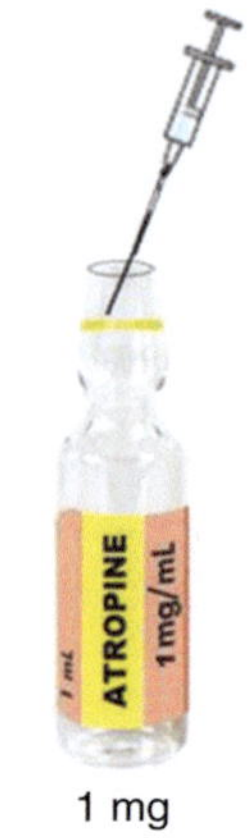

Fig. 19.17 Atropine dose in children (Sledz's study)

cases of PR/RR > 1. Authors conclude that PR/RR ration >1 is a better sign of slow pathway presence than the conduction jump. Finally, the absence of PR/RR interval >1 after catheter ablation is a better marker of slow pathway ablation than the absence of jump >40–50 ms [23].

19.4 Adenosine for Arrhythmia Induction in Children

Adenosine is a highly effective medicine which is mostly used for SVT termination. In children doses are in range from 50 to 350 μg/kg. Starting dose of 50–100 μg/kg is successful in at most one third of patients. When unsuccessful, every next dose, which is given after 3–4-min pause, is raised for 50 μg/kg. Maximal dose is 12 mg. There are centers that use the starting dose of 100–150 μg/kg, and the next doses are 200 and 300 μg/kg [24–30].

In the study of Curtis et al., low-dose adenosine was injected for AVNRT induction. The fast and slow AV node pathways do not have similar response to adenosine; the modification of refractory periods can facilitate reentry. After small doses of adenosine, antegrade fast pathway is more sensitive than the slow pathway, and adenosine blocks the fast pathway, facilitating conduction through the slow pathway and reentry [31]. Curtis et al. injected adenosine in progressive increasing doses, 1, 2, 4, 6, 9, 12, and 18 mg in 26 patients during sinus rhythm, to initiate AVNRT. The dose that blocked the fast pathway and facilitated AVNRT was 3.2 mg. Doses of 50–350 μg/kg/min are used in children to stop a paroxysmal supraventricular tachycardia, with a maximum dose of 12 mg. As an extrapolation, using the Johnson formula, 10–80 μg/kg could be sufficient to induce AVNRT in children, but this observation needs further studies to confirm it (Fig. 19.18) [32].

In the study of Toal et al., 97 patients received 12 mg of adenosine during narrow QRS paroxysmal supraventricular tachycardia to differentiate between AVNRT and orthodromic reentrant atrioventricular tachycardia (AVRT). Patients were both children and adults, and the youngest patient was 12 years. Some of the patients presented during adenosine test a PR jump and atrioventricular node reentrant

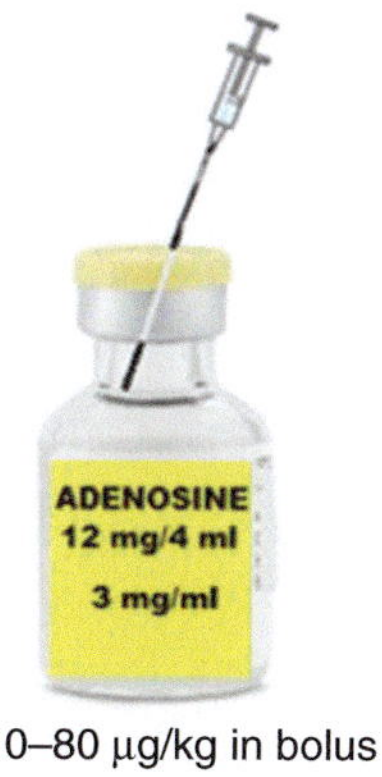

Fig. 19.18 Adenosine dose in children

tachycardia and others AVRT using an accessory pathway. Therefore, adenosine can be used not only to stop a paroxysmal reentrant tachycardia but also to initiate it because it modifies both the refractory period of the slow pathway and fast nodal pathway.

In the EP laboratory, adenosine can also be injected to unreveal conduction through an accessory pathway—which is absent at baseline, after the mechanical block or after an apparently successful ablation. In this setting adenosine should be injected in doses of 200–300 μg/kg (max 12 mg). In the biggest series on six children, the dose of adenosine was 100–200 μg/kg [33].

References

1. Gautam NK, Rafique MB, Numan MT. Impact of isoproterenol infusion on BIS and metabolic values in pediatric patients undergoing electrophysiology studies. J Clin Anesth. 2014;26:611–5.
2. Moore JP, Kannankeril PJ, Fish FA. Isoproterenol administration during general anesthesia for the evaluation of children with ventricular preexcitation. Circ Arrhythm Electrophysiol. 2011;4:73–8.
3. Isuprel (isoproterenol) [prescribing information]. Lake Forest, IL: Hospira, Inc.; 2013.
4. Kubuš P, Vít P, Gebauer RA, Materna O, Janoušek J. Electrophysiologic profile and results of invasive risk stratification in asymptomatic children and adolescents with the Wolff-Parkinson-white electrocardiographic pattern. Circ Arrhythm Electrophysiol. 2014;7:218–23.
5. Liberman L, Hordof AJ, Fishberger SB, Pass RH. The role of isoproterenol testing following radiofrequency catheter ablation of accessory pathways in children. Pacing Clin Electrophysiol. 2003;26:559–61.
6. Brembilla-Perrot B, Brembilla A, Moulin-Zinsch A, Sellal JM, Olivier A, Manenti V, et al. Factors of negativity of electrophysiological study in children and teenagers complaining of tachycardia and prognostic significance. Pediatr Cardiol. 2015;36:64–70.
7. Mah DY, Sherwin ED, Alexander ME, Cecchin F, Abrams DJ, Walsh EP, et al. The electrophysiological characteristics of accessory pathways in pediatric patients with intermittent preexcitation. Pacing Clin Electrophysiol. 2013;36:1117–22.
8. Brembilla-Perrot B, Pauriah M, Sellal JM, Zinzius PY, Schwartz J, de Chillou C, et al. Incidence and prognostic significance of spontaneous and inducible antidromic tachycardia. Europace. 2013;15:871–6.

9. Mills MF, Motonaga KS, Trela A, Dubin AM, Avasarala K, Ceresnak SR. Is there a difference in tachycardia cycle length during SVT in children with AVRT and AVNRT? Pacing Clin Electrophysiol. 2016;39:1206–12.
10. Emmel M, Sreeram N, Brockmeier K. Catheter ablation of junctional ectopic tachycardia in children, with preservation of atrioventricular conduction. Z Kardiol. 2005;94:280–6.
11. Brembilla-Perrot B, Moulin-Zinsch A, Sellal JM, Schwartz J, Olivier A, Zinzius PY, et al. Impact of transesophageal electrophysiologic study to elucidate the mechanism of arrhythmia on children with supraventricular tachycardia and no preexcitation. Pediatr Cardiol. 2013;34:1695–702.
12. Drago F, De Santis A, Grutter G, Silvetti MS. Transvenous cryothermal catheter ablation of re-entry circuit located near the atrioventricular junction in pediatric patients: efficacy, safety, and midterm follow-up. J Am Coll Cardiol. 2005;45:1096–103.
13. Ko JK, Ryu SJ, Ban JE, Kim YH, Park IS. Use of transesophageal atrial pacing for documentation of arrhythmias suspected in infants and children. Jpn Heart J. 2004;45:63–72.
14. Law IH, Fischbach PS, Goldberg C, Mosca RS, Bove EL, Lloyd TR, et al. Inducibility of intra-atrial reentrant tachycardia after the first two stages of the Fontan sequence. J Am Coll Cardiol. 2001;37:231–7.
15. Shih HT, Miles WM, Klein LS, Hubbard JE, Zipes DP. Multiple accessory pathways in the permanent form of junctional reciprocating tachycardia. Am J Cardiol. 1994;73:361–7.
16. Kleinman ME, Chameides L, Schexnayder SM, Samson RA, Hazinski MF, Atkins DL, et al. Part 14: pediatric advanced life support: 2010 American Heart Association Guidelines for Cardiopulmonary Resuscitation and Emergency Cardiovascular Care. Circulation. 2010;122:S876–908.
17. American Heart Association. 2005 American Heart Association (AHA) guidelines for cardiopulmonary resuscitation (CPR) and emergency cardiovascular care (ECC) of pediatric and neonatal patients: pediatric advanced life support. Pediatrics. 2006;117:e1005–28.
18. Jones P, Peters MJ, Pinto da Costa N, Kurth T, Alberti C, Kessous K, et al. Atropine for critical care intubation in a cohort of 264 children and reduced mortality unrelated to effects on bradycardia. PLoS One. 2013;8:e57478.
19. Pihlajamäki K, Kanto J, Aaltonen L, Iisalo E, Jaakkola P. Pharmacokinetics of atropine in children. Int J Clin Pharmacol Ther Toxicol. 1986;24:236–9.
20. Unna KR, Glaser K, Lipton E, Patterson PR. Dosage of drugs in infants and children. I. Atropine. Pediatrics. 1950;6:197–207.
21. Marcus B, Gillette PC, Garson A Jr. Electrophysiologic evaluation of sinus node dysfunction in postoperative children and young adults utilizing combined autonomic blockade. Clin Cardiol. 1991;14:33–40.
22. Sileikiene R, Baksiene D, Sileikis V, Kazakavicius T, Vaskelyte J, Kevalas R. Changes in electrophysiologic properties of the conductive system of the heart in children with atrioventricular nodal reentrant tachycardia after 2-8 years following radiofrequency catheter ablation of the slow pathway. Medicina (Kaunas). 2009;45:632–8.
23. Sledz J, Labus M, Mazij M. A simplified approach for evaluating sustained slow pathway conduction for diagnosis and treatment of atrioventricular nodal reentry tachycardia in children and adults. Adv Med Sci. 2018;63:249–56.
24. Qureshi AU, Hyder SN, Sheikh AM, Sadiq M. Optimal dose of adenosine effective for supraventricular tachycardia in children. J Coll Physicians Surg Pak. 2012;22:648–51.
25. Sherwood MC, Lau KC, Sholler GF. Adenosine in the management of supraventricular tachycardia in children. J Paediatr Child Health. 1998;34:53–6.
26. Díaz-Parra S, Sánchez-Yañez P, Zabala-Argüelles I, Picazo-Angelin B, Conejo-Muñoz L, Cuenca-Peiró V, et al. Use of adenosine in the treatment of supraventricular tachycardia in a pediatric emergency department. Pediatr Emerg Care. 2014;30:388–93.
27. Till J, Shinebourne EA, Rigby ML, Clarke B, Ward DE, Rowland E. Efficacy and safety of adenosine in the treatment of supraventricular tachycardia in infants and children. Br Heart J. 1989;62:204–11.

28. Losek JD, Endom E, Dietrich A, Stewart G, Zempsky W, Smith K. Adenosine and pediatric supraventricular tachycardia in the emergency department: multicenter study and review. Ann Emerg Med. 1999;33:185–91.
29. Dixon J, Foster K, Wyllie J, Wren C. Guidelines and adenosine dosing in supraventricular tachycardia. Arch Dis Child. 2005;90:1190–1.
30. Paul T, Pfammatter JP. Adenosine: an effective and safe antiarrhythmic drug in pediatrics. Pediatr Cardiol. 1997;18:118–26.
31. Curtis AB, Belardinelli L, Woodart DA, Rown CS, Conti JB. Induction of atrioventricular node reentrant tachycardia with adenosine: differential effect of adenosine on fast and slow atrioventricular node pathways. JACC. 1997;30:1778–84.
32. Johnson TN. The problem in scaling adult drug doses in children. Arch Dis Child. 2008;93:207–11.
33. Lapage MJ, Walsh MJ, Reed JH, Saul JP. Adenosine mapping for adenosine-dependent accessory pathway ablation. Pacing Clin Electrophysiol. 2014;37:610–5.

Arrhythmia Induction During Transesophageal Electrophysiological Study

20

Keith Andrew Chan

20.1 Introduction to Transesophageal Electrophysiological Study

Electrophysiologic studies are a series of invasive diagnostic procedures performed in a dedicated electrophysiology suite or laboratory. The main purposes of these studies are to establish the presence of various forms of arrhythmias, to establish baseline information in patients who are already receiving treatment or are due to undergo pharmagologic, electrical, or ablative treatment for arrhythmias and as workup for various electrophysiologic causes of chief complaints such as syncope and cardiac arrest. Such studies may also be undertaken in children with various forms of congenital heart disease before and after they have undergone reparative surgery [1, 2]. With this, EP studies remain an indispensable part of today's modern heart institutes and continue to remain a field of rapid advancement and refinement.

In addition to the various catheter-based diagnostic methods used in the catheterization laboratory, the transesophageal electrocardiogram was developed in the 1950s as a means for more accurate arrhythmia recognition and identification [3] This tool was developed on the premise of the esophagus' closer proximity to the posterior wall of the left atrium, making it an attractive tool for more accurate and reliable cardiac pacing and rhythm studies [3]. Although relatively inexpensive compared to catheter-based studies and despite the ability to perform transesophageal electrophysiologic studies (TEEPS) without specialized fluoroscopy equipment, its use has generally remained limited to a handful of clinical indications, notably among the pediatric population. No large society consensus guidelines presently exist for the main indications of TEEPS, but they are generally performed for provocation of previously non-registered symptomatic supraventricular tachycardias (SVT) [3]. Additional but lesser studied indications include sinus and

K. A. Chan (✉)
Section of Adult Cardiology, Chong Hua Heart Institute, Chong Hua Hospital, Cebu, Philippines

G. Cismaru (ed.), *Arrhythmia Induction in the EP Lab*, https://doi.org/10.1007/978-3-319-92729-9_20

atrioventricular (AV) node evaluation and assessment of palpitations of unknown origin [3–5]. Electrophysiologic interventions may also be undertaken by means of transesophageal pacing (TEAP) for the termination of AVRT and AVNRT [6]. A recent large-scale database review done in Ankara, Turkey, revealed risk assessment for Wolff-Parkinson-White syndrome and palpitation etiology determination to be the major indications for the performance of TEEPS [7].

20.2 Technique of Transesophageal Electrophysiological Study

TEEPS is a noninvasive electrophysiologic study that can be performed in inpatient settings or outpatient clinics. TEPS is a highly convinient and economically feasible diagnotic tool that requires minimal training and sedation to perform. However, the author notes the absence of widely circulated competency checklists or procedural blueprints for TEPS [3, 7–9]. A brief procedure description for the performance of TEEPS is described.

20.2.1 Patient Preparation and Sedation

Patient preparation has been described to be relatively straightforward: Pehrson's protocol (1994) describes a need to discontinue the patient's anti-arrhythmic medications for at least five half-lives prior to the scheduled procedure [10]. Patients are then required to fast for a period of at least 4 h [5]. Intravenous access and line patency are ensured prior to the performance of the procedure. A variety of sedation methods have been used at various centers, particularly on the pediatric population for which this procedure is often recommended. Brembilla-Perrot (2009) initially described a protocol only requiring oral cavity anesthesia in a patient with exaggerated nausea prior to the procedure [8]. However, a database review by Gilly et al. (2015) revealed much more aggressive approaches to sedation in various attempts to yield reliable and reproducible results for both electrophysiologists and anesthesiologists with minimal patient discomfort. The best regimen, as recommended by Gilly, was a combination of mask sedation with continuous intravenous (IV) sedation [11]. Patients are positioned in a semi-recumbent position with easy access for nasal or oral access.

20.2.2 Electrode Insertion and Placement

TEEPS utilizes a wide range of flexible silicone-coated unipolar and bipolar electrode-containing catheters (Fig. 20.1). For the more frequently used bipolar variants, various interelectrode distances have been described, ranging from 15 to 30 mm with wider distances generally reserved for older patients with larger habitus [12, 13]. No distinct protocol or anatomic landmarks have been described as a basis

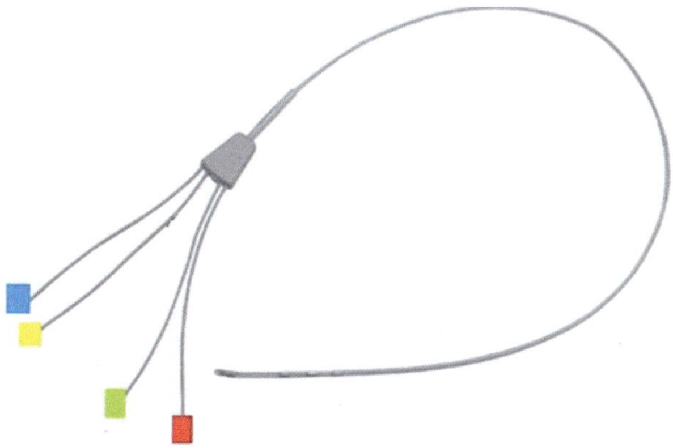

Fig. 20.1 A bipolar pacing and recording catheter

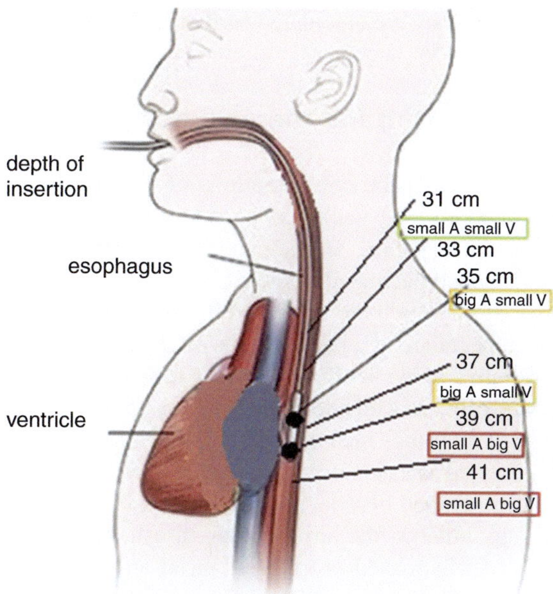

Fig. 20.2 Diagram showing the varying esophageal electrode insertion depths and their relationships to cardiac structures and corresponding electrocardiograms

for interelectrode distance selection, but physician discretion and variability in choice of distances have shown to have no significant effect on procedure performance [14]. The catheter is lubricated and slowly advanced into the esophagus either through the nares or the oral cavity [10, 13]. The depth of insertion (DOI) for optimum atrial pacing has been described to be at a mean of approximately 40 cm from the nares and oral cavity (Fig. 20.2). However, given the variability of patient age and size, landmarks are generally required to better establish an optimum point. Roth et al. (1996) describes a DOI from the oral cavity of approximately one-half the patient's height in centimeters with an optimum atrial capture established at approximately ±3 cm of this product, with an additional 3–4 cm added if nasal access is used [14]. Once optimum electrode placement has been established, it is then secured in place.

20.2.3 Baseline Electrocardiography and Recording of Arrhythmias

Before pacing or other therapeutic interventions, the patient is attached to a standard surface 12-lead electrocardiogram (ECG) with an additional esophageal lead. The esophageal lead, being the closest to the left atrium, manifests with larger P wave complexes, sometimes uncovering obscured P waves seen on surface ECG recordings, as was first demonstrated by Brown in 1936 [15]. A combination of the esophageal lead recording with or without a superimposed 12-lead surface ECG is taken and subsequently analyzed. The principles of esophageal lead recording and wave analysis are elucidated further in the next session.

20.2.4 Pacing and Therapeutic Maneuvers

Depending on the considerations of the electrophysiologic study or the need for therapeutic TAP, pacing is carried out in a manner similar to transcutaneous or transvenous pacing. Based on published experience with various pacing modalities, the attached pacing equipment's stimulator enables a present rate, duration, and current, all of which are increased in a stepwise manner. Capture is then confirmed by the appearance of a pacing spike on the patient's surface and esophageal electrocardiogram tracing with a 1:1 pacer to QRS relationship noted. However, there are several special considerations for the transesophageal approach to pacing. Benson (1984) describes the concept of a "pacing window" or a specific esophageal region (expressed as a distance from the nares or oral cavity) over which a current of 20 mA or less resulted in atrial pacing. Benson further describes a relationship between the pacing window, the atrial pacing threshold, and the esophageal ECG atrial amplitude. An increased esophageal atrial amplitude was noted at sites of lower atrial pacing thresholds which in turn have roughly a 1.1–3 cm proximity to the atrial pacing window. Thus, placement of either a unipolar or bipolar electrode over this window ensures maximum atrial pacing with minimal electrical stimuli [12]. As previously described, Roth's method produced reliable and acceptable TAP in study patients, but the author notes that large-scale validation studies are lacking to confirm its efficacy [14].

20.2.5 Procedure Termination and Aftercare

Once TAP and other diagnostic procedures have finished, the machine's pacing functions are turned off and ensured that no electrical current is present within the system. The electrodes are removed, and postanesthetic care is rendered to the patient based on the type of anesthesia administered prior and during the procedure.

20.3 Principles of Electrocardiography in TEEPS

As previously mentioned, TEEPS utilizes an additional transesophageal electrode positioned in a region assumed to be closest to the left atrium where both maximum P wave amplitude and pacing capabilities are manifested. The fundamental underlying principle of TEEPS that makes it highly useful in clinical practice is its ability to record atrial depolarization despite its absence on surface electrocardiography, making it useful in cases of blocked premature atrial contractions (PACs) overlapped by T waves, second-degree AV blocks, SVTs with aberrant conduction, and workup for wide complex tachycardias [16–18]. The increased sensitivity of TEEPS in the detection of atrial activity is best summed up by a series of observations made by Jenkins and Arzbaecher (1984): (a) Chest electrodes applied closest to the sternum have lead fields on average only one-fourth as strong as those of ventricular surfaces. (b) Unipolar esophageal electrodes have lead fields at the left atrium eight times stronger than that of ventricular surfaces, while bipolar leads have almost 20 times stronger fields for the left atrium compared to ventricles. (c) Both unipolar and bipolar leads also have two and a half times greater lead fields for the His bundle compared to surface or precordial electrodes [19]. Given its superiority in its assessment of the P wave and, thus, atrial rhythms, it comes to no surprise that more complex cardiac problems have required esophageal lead analysis for more accurate diagnosis of arrhythmias [20].

20.4 Mechanisms of Arrhythmia Induction

Among the various capabilities of transesophageal electrode systems, the author highlights its ability to induce several distinct arrhythmias, namely, atrial fibrillation and flutter and various other forms of supraventricular tachycardias (SVTs). The specific methods of induction and clinical uses are mentioned in each specific subcategory of arrhythmias below. The main principle of arrhythmogenesis of these SVTs lies in the ability of transesophageal electrodes to reliably and consistently induce atrial pacing, which in turn stimulates or excites atrial tissue. Arrhythmias are thus produced in susceptible individuals in whom a critical mass of atrial tissue exhibits electrophysiologic properties with sustained reentry mechanisms [21, 22]. However, despite all the convenience that TEEP offers, we also have to consider its limitations in arrhythmia assessment.

20.4.1 Atrial Fibrillation and Flutter

The induction of atrial fibrillation and/or flutter is done for two main indications: Kerr (1983) described the utility of esophageal electrode-based induction of atrial fibrillation and flutter as part of the overall EP assessment of Wolff-Parkinson-White (WPW) patients, as spontaneous occurrences of these tachyarrhythmias may prove to be fatal in its induction of hemodynamic compromise [23]. Another clinical use of atrial fibrillation/flutter induction was described by Brembilla-Perrot (2015) as a means identifying patients at high risk of recurrent atrial fibrillation/flutter not identified on noninvasive monitoring methods [24]. The specific methods with which atrial fibrillation/flutter was induced were elucidated more explicitly in Kerr's article (Table 20.1): a bipolar esophageal electrode with a 29-mm interelectrode distance was positioned at the area of the largest and most rapid electrocardiogram (mean distance of 36.6 ± 2.9 cm from nares). Pacing was subsequently done with the listed parameters in Table 20.1. Out of the 23 WPW patients assessed, 22 patients had either atrial fibrillation, flutter, or flutter/fibrillation successfully induced [23]. Atrial fibrillation can be induced by fast atrial fibrillation (Figs. 20.3 and 20.4).

Table 20.1 Atrial fibrillation/flutter induction methodology (Kerr) [23]

Population: Patients with documented WPW syndrome
Electrode: Bipolar, esophageal; 29 mm inter electrode distance
Position: 36.6 + 2.9 cm from nares
Pacing: • Cycle length: 40–340 ms (mean 166 + 72 ms) • Pulse duration: 7.0–9.9 ms • Current: 10–25 mA
Termination: Esophageal pacing at cycle of 160–220 ms (mean 176 ± 18 ms)
Induction rate: 96% (22 out of 23 patients)

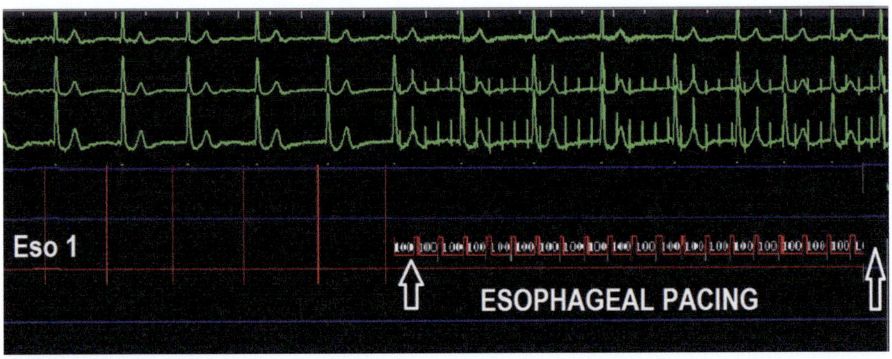

Fig. 20.3 Fast atrial fibrillation. Stimulation is performed in the esophagus at 35 cm from the lips which permits capture of the left atrium

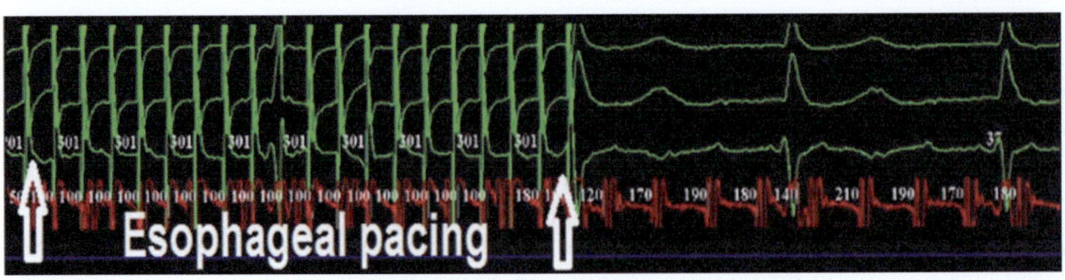

Fig. 20.4 Fast esophageal pacing induces atrial fibrillation

20.4.2 AVNRT and AVRT

One of the main clinical indications for the performance of TEAP is its ability to noninvasively induced SVTs in patients with either ECG-documented paroxysmal SVT or with a clinical history compatible with SVT [10]. Pehrson (1994) elegantly described a stepwise protocol for the induction of reentrant SVTs (summarized in Table 20.2): a bipolar electrode (interelectrode distance of 15 mm) was passed via the nares into a position of maximum ECG amplitude. Atrial pacing was then performed via a programmable stimulator and a pulse generator with a fixed pulse duration of 10 ms. Incremental atrial stimulation was then done at 10 ms below the basic cycle length with stepwise decrements of 25 ms to either a basic cycle of 275 ms or until a second-degree AV block appeared. Further steps for induction of SVTs were done by double extrastimuli delivered at a drive cycle of 50 ms above the refractory period until refractoriness is achieved. IV atropine was given as the last final step for the induction of SVTs. Out of 39 patients, 35 patients had SVT successfully induced via the study's esophageal electrode [10]. Pehrson further describes that a 90% induction rate is for patients with documented paroxysmal SVTs but only a 73% induction rate for patients undergoing workup for palpitations [10]. The specific SVTs induced in these patients were a mix of either AVNRT or AVRT which correlated with their previously seen PSVT documented on their baseline ECG. For patients without baseline preexcitation recorded on their ECG, SVTs were still induced, albeit at a lower rate. A similar protocol for the pediatric population was described by Drago (1996) instead of utilizing quadripolar esophageal electrodes, although only distal electrodes were used for pacing. Drago's protocol also incorporated stress testing on top of transesophageal atrial pacing in order to better assess the occurrence of SVTs in relation to the effort in the investigation of SVT among the pediatric population [25].

Table 20.2 Supraventricular tachycardia (SVT) induction (Pehrson) [10]

Population: Patients with documented paroxysmal SVT (*n* = 39) or complaints of palpitations with unknown etiology (*n* = 11)
Electrode: Bipolar, esophageal; 15 mm interelectrode distance
Position: Unspecified; maximum esophageal ECG amplitude
Pacing: • **Initiation**: 10 ms below cycle length; decreased by 25 ms decrement to 275 ms OR appearance of second-degree AV block • **Second phase**: Double extrastimuli at drive cycle of 500 ms with 10 ms decrements starting at 50 ms above refractory period • **Third phase**: Atropine IV (0.02 mg/kg)
Termination: Overdrive atrial pacing done; specifications not given
Induction rate:
90% (35 of 39 patients) with documented paroxysmal SVT
73% (8 of11 patients) with palpitations of unknown etiology

20.4.3 Atrial Tachycardia and Premature Junctional Tachycardia

Literature on the induction of SVTs secondary to altered automaticity remains sparse. However, a study performed by Brembilla-Perrot (1990) describes a protocol with incremental atrial stimulation delivering one and two extrastimuli on two paced rhythms (400–600 ms) under basal conditions and after the infusion of isoproterenol. Out of the two patient populations, it was noted to be far more difficult to induce atrial tachycardia compared to premature junctional tachycardia (100% induction rate in PJCs vs. 74% induction rate for AT) [26].

20.5 Summary

In summary, transesophageal electrophysiologic studies remain a useful but underutilized modality in the electrophysiology suite. It is a safe and semi-invasive procedure that can be performed with minimal training and laboratory equipment. As of the time of writing, its main clinical utility lies in the assessment of WPW syndrome patients and as part of the workup regimen for various SVT etiologies as well as hemodynamic risk assessment for patients with known WPW. The procedure remains relatively safe and simple, utilizing endoscopic electrode advanced under local or systemic sedation. With current generation electrodes, the present literature describes several feasible methods used for the induction of atrial fibrillation, atrial flutter, and various supraventricular tachycardias in the EP laboratory.

References

1. Guidelines for Clinical Intracardiac Electrophysiologic Studies. A report of the American College of Cardiology/American Heart Association Task Force on Assessment of Diagnostic and Therapeutic Cardiovascular Procedures (Subcommittee to Assess Clinical Intracardiac Electrophysiologic Studies). Circulation. 1989;80:1925–39.
2. Zipes DP, Dimarco JP, Gillette PC, Jackman WM, Myerburg RJ, Rahimtoola SH, Ritchie JL, Cheitlin MD, Garson A, Gibbons RJ, Lewis RP, O'Rourke RA, Ryan TJ, Schlant RC. Guidelines for clinical intracardiac electrophysiological and catheter ablation procedures. J Am Coll Cardiol. 1995;26:555–73.
3. Szili-Torok T, Mikhaylov E, Witsenburg M. Transoesophageal electrophysiology study for children: can we swallow the limitations? Europace. 2009;11:987–8.
4. Gallagher JJ, Smith WM, Kerr CR, Kasell J, Cook L, Reiter M, Sterba R, Harte M. Esophageal pacing: a diagnostic and therapeutic tool. Circulation. 1982;65:336–41.
5. Verbeet T, Castro J, Decoodt P. Transesophageal pacing: a versatile diagnostic and therapeutic tool. Indian Pacing Electrophysiol J. 2003;3:202–9.
6. Volkmann H, Dannberg G, Heinke M, Kühnert H. Termination of tachycardias by transesophageal electrical pacing. Pacing Clin Electrophysiol. 1992;15:1962–6.
7. Erdoğan I1, Ozer S, Karagöz T, Sahin M, Celiker A. Clinical importance of transesophageal electrophysiologic study in the management of supraventricular tachycardia in children. Turk J Pediatr. 2009;51:578–81.
8. Brembilla-Perrot B, Groben L, et al. Rapid and low-cost method to prove the nature of no documented tachycardia in children and teenagers without pre-excitation syndrome. Europace. 2009;11:1083–9.

9. Brembilla-Perrot B, Chometon F, et al. Interest of non-invasive and semi-invasive testings in asymptomatic children with pre-excitation syndrome. Europace. 2007;9:837–43.
10. Pehrson SM, Blomstrom-Lundqvist C, Ljungstrom E, Blomstrom P. Clinical value of transesophageal atrial stimulation and recording in patients with arrhythmia-related symptoms or documented supraventricular tachycardia correlation to clinical history and invasive studies. Clin Cardiol. 1994;17:528–34.
11. Gilly G, Hoyt W, Harmon D, Busch E, Nossaman B, Broussard D, Snyder C. Anesthetic technique for transesophageal electrophysiology studies in pediatric patients with Wolff-Parkinson-White syndrome. Open J Pediatr. 2015;5:17–22.
12. Benson DW Jr, Sanford M, Dunnigan A, Benditt DG. Transesophageal atrial pacing threshold: role of interelectrode spacing, pulse width and catheter insertion depth. Am J Cardiol. 1984;53:63–7.
13. FIAB. Tranesophageal and diagnostic pacing leads specifications. http://www.fiab.it/en/prodotti.php?id=36.
14. Roth JV, Brody JD, Denham EJ. Positioning the pacing esophageal stethoscope for transesophageal atrial pacing without P-wave recording: implications for transesophageal ventricular pacing. Anesth Analg. 1996;83:48–54.
15. Brown WH. A study of the esophageal lead in clinical electrocardiography. Am Heart J. 1936;12:307–38.
16. Butterworth S, Poindexter CA. The esophageal electrocardiogram in arrhythmias and tachycardias. Am Heart J. 1946;32:681–8.
17. Enselberg CD. The esophageal electrocardiogram in the study of atrial activity and cardiac arrhythmias. Am Heart J. 1951;41:382.
18. Tiwari VD, Balasubramanian V, Durairaj M, Hoon RS. The esophageal electrocardiogram in normal Indian subjects. Indian J Med Res. 1975;63:422.
19. Jenkins JM, Arzbaecher R. Chapter 14: esophageal electrocardiography. In: Liebman J, Plonsey R, Rudy Y, editors. Pediatric and fundamental electrocardiography. New York: Springer; 1987.
20. Bushman GA. Clinical correlates of dysrhythmias requiring an esophageal ECG for accurate diagnosis in patients with congenital heart disease. J Cardiothorac Anesth. 1989;3:290–4.
21. Brignole M, et al. The use of atrial pacing to induce atrial fibrillation and flutter. Int J Cardiol. 1986;12:45–54.
22. Gallagher JJ, Smith WM, et al. Esophageal pacing: a diagnostic and therapeutic tool. Circulation. 1982;652:336–41.
23. Kerr CR, Gallagher JJ, Smith WM, et al. The induction of atrial flutter and fibrillation and the termination of atrial flutter by esophageal pacing. Pacing Clin Electrophysiol. 1983;6:60–72.
24. Brembilla-Perrot B, Olivier A, Girerd N, Manenti V, Sellal JM, Rizk J, Villemin T. 0019: significance of atrial fibrillation/tachycardia induced by esophageal stimulation. Arch Cardiovasc Dis Suppl. 2015;7:72.
25. Drago F, Turchetta A, Calzolari A, Giordano U, Di Ciommo V, Santilli A, et al. Reciprocating supraventricular tachycardia in children: low rate at rest as a major factor related to propensity to syncope during exercise. Am Heart J. 1996;132:280–5.
26. Brembilla-Perrot B, Spatz F, Khaldi E, Terrier de la Chaise A, Suty-Selton C, Le Van D, Cherrier F, Pernot C. Induction of supraventricular tachycardia (paroxysmal junctional tachycardia and atrial tachycardia) by esophageal stimulation. Arch Mal Coeur Vaiss. 1990;83:1695–702.

21 How to Induce Pacemaker-Mediated Tachycardia?

Gusetu Gabriel, Lucian Muresan, Rosu Radu, Horatiu Comsa, Caloian Bogdan, Dana Pop, and Dumitru Zdrenghea

21.1 Introduction

Pacemakers are devices that are used worldwide for bradyarrhythmias. They are robust devices that function properly from implantation to depletion of battery. Rarely malfunctions appear, one of them being pacemaker-mediated tachycardia in patients implanted with double-chamber devices when retrograde ventriculoatrial conduction is present. Retrograde ventriculoatrial conduction can be present in up

G. Gabriel (✉) · R. Radu · H. Comsa · C. Bogdan · D. Pop · D. Zdrenghea
"Iuliu Hatieganu University of Medicine and Pharmacy", Rehabilitation Hospital, Cluj-Napoca, Romania
e-mail: gusetu.gabriel@umfcluj.ro; rosu.radu@umfcluj.ro; comsa.horatiu@umfcluj.ro; caloian.bogdan@umfcluj.ro; pop.dana@umfcluj.ro; zdrenghea.dumitru@umfcluj.ro

L. Muresan
Cardiology Department, Hopital "Emile Muller", Mulhouse, France
e-mail: muresan.lucian@umfcluj.ro

G. Cismaru (ed.), *Arrhythmia Induction in the EP Lab*,
https://doi.org/10.1007/978-3-319-92729-9_21

to 50–70% of normal patients [1–3]. In patients with a normal heart, usually it does not manifest clinically. Even in the case of an advanced AV block, retrograde VA conduction can still be present. Patients implanted with a cardiac device like pacemakers or defibrillators can exhibit ventriculoatrial conduction: 14% of patients with complete AV block and up to 80% of patients with sinus node disease [4]. In patients with double-chamber pacemakers or double-chamber defibrillators, using an atrial sensing mode can serve as a second AV pathway and support reentry: antegrade pathway through the pacemaker and retrograde pathway through the AV node.

Ventriculoatrial conduction has been frequently implicated as a possible cause of the "pacemaker syndrome" in patients with double-chamber devices and endless loop tachycardia and detection issues in patients implanted with double-chamber defibrillators. If a retrograde P wave is conducted from the ventricles to the atria, it is sensed by the atrial channel in double-chamber pacemakers and can induce a form of pacemaker-mediated tachycardia called endless loop tachycardia. Despite development of several algorithms for detection and termination of this type of pacemaker-induced tachycardia, it still remains a problem that should be overcome because of symptoms and impact on cardiac function as it is responsible for palpitation, dyspnea, heart failure, and other cardiac symptoms. ICDs are also very sensible to retrograde conduction. Algorithms are made to differentiate between supraventricular and ventricular tachycardia. When a tachycardia has 1:1 retrograde conduction, the device will classify it as supraventricular tachycardia. When the QRS is wide, another algorithm of the device will consider it ventricular tachycardia. Problems appear in the case of ventricular tachycardia with 1:1 conduction to the atria because defibrillator algorithms could consider it supraventricular tachycardia and not deliver overdrive pacing or internal electrical shock.

Some patients can present ventriculoatrial conduction during physiological or pathological stresses when the concentration of catecholamines increases in the body. Several medications were demonstrated to induce ventriculoatrial retrograde conduction in patients with absent conduction in the basal state. Adrenaline was demonstrated to reverse absent ventriculoatrial conduction, a phenomenon that could explain the presence of pacemaker syndrome, endless loop tachycardia, or even detection discrimination troubles in ICD in patients without ventriculoatrial conduction in the basal state.

21.2 Induction of Retrograde VA Conduction with Atropine

Neural influences on the AV node are both vagal and sympathetic. Daily activities act both on sympathetic and vagal tones, and appropriate autonomic tone influences on the AV node could be tested by using adrenaline for the sympathetic system and atropine for the parasympathetic-vagal system. Sadr-Ameli et al. used carotid sinus pressure as surrogate of vagal stimulation, propranolol for sympathetic inhibition, atropine for vagal inhibition, and isoproterenol as sympathetic stimulator. They have also demonstrated on 16 patients with intact antegrade AV conduction and absent retrograde VA conduction that retrograde VA conduction can be induced

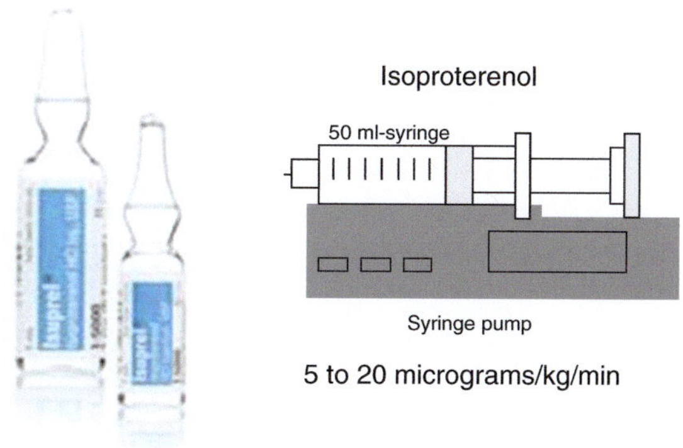

Fig. 21.1 Isoprenaline infusion for induction of ventriculoatrial conduction

with either atropine or isoprenaline: eight patients received 0.04 mg/kg of atropine, and eight patients received isoproterenol in doses of 1 μg/min. Atropine induced retrograde VA conduction in six of the eight patients (75%), and isoproterenol induced retrograde VA conduction in six of the eight patients (75%). They showed a similar inducibility rate for both atropine and isoprenaline [5].

Hariman et al. also showed that isoproterenol and atropine can induce ventriculoatrial conduction. They infused isoproterenol and injected atropine in 17 patients without VA conduction. Five of the 17 patients (29.4%) showed induction of VA conduction after isoproterenol infusion and 6 of the 17 (35.3%) induction of VA conduction after atropine injection [6] (Fig. 21.1).

21.3 Induction of Retrograde VA Conduction with Isoprenaline

As mentioned above, both atropine and isoprenaline can induce retrograde ventriculoatrial conduction in patients that have absent conduction in the basal state. Sadr-Ameli showed on eight patients who received 1 μg/min isoprenaline that it can facilitate VA conduction when it is absent in the basal state. Also Hariman et al. showed on 17 patients that after induction, retrograde VA conduction was present in 6 of the patients (Fig. 21.2).

21.4 Induction of Retrograde VA Conduction with Adrenaline

Adrenaline shortens the effective refractory period of the atrium, AV node, and ventricle and also improves the electrical conduction through the AV node [7]. A physiological increase in the circulating adrenaline can shorten atrial and atrioventricular nodal refractoriness and accelerate AV nodal electrical conduction as well as

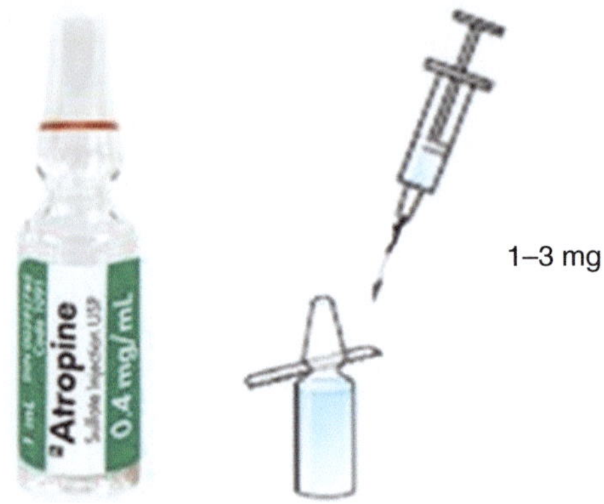

Fig. 21.2 Atropine injection for induction of ventriculoatrial conduction

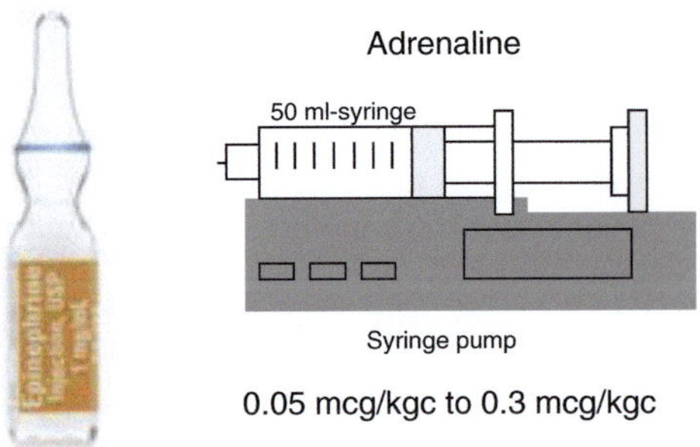

Fig. 21.3 Adrenaline infusion for induction of ventriculoatrial conduction

decrease of the Wenckebach point (longest pacing cycle length associated with AV block) [8].

Studies demonstrated that 0.025–0.050 μg/kg/min results in plasma adrenaline concentrations comparable with those concentrations that occur during various physiological or pathophysiological stresses: the infusion of a dose of 0.025 μg/kg/min produces a plasma concentration observed during cigarette smoking, public speaking, submaximal exercise, mild hypoglycemia, dental extraction, or surgery [9]. Infusion of 0.05 μg/kg/min adrenaline results in plasma concentrations observed during maximal exercise, myocardial infarction, diabetic ketoacidosis, and severe hypoglycemia [9] (Fig. 21.3).

Cismaru et al. demonstrated that adrenaline can be used for ventriculoatrial induction in patients with absent retrograde VA conduction in the basal state [9]. They infused adrenaline in 46 patients with absent VA conduction and recovered the VA conduction in 5 of them (10.9%) (Fig. 21.4). Patients were implanted with a Medtronic or Saint Jude device, had documented second-degree or third-degree AV block, or had sinus node disease. The mean dose of adrenaline was 0.1 μg/kg/min. All of the patients showed increase in heart rate after adrenaline infusion

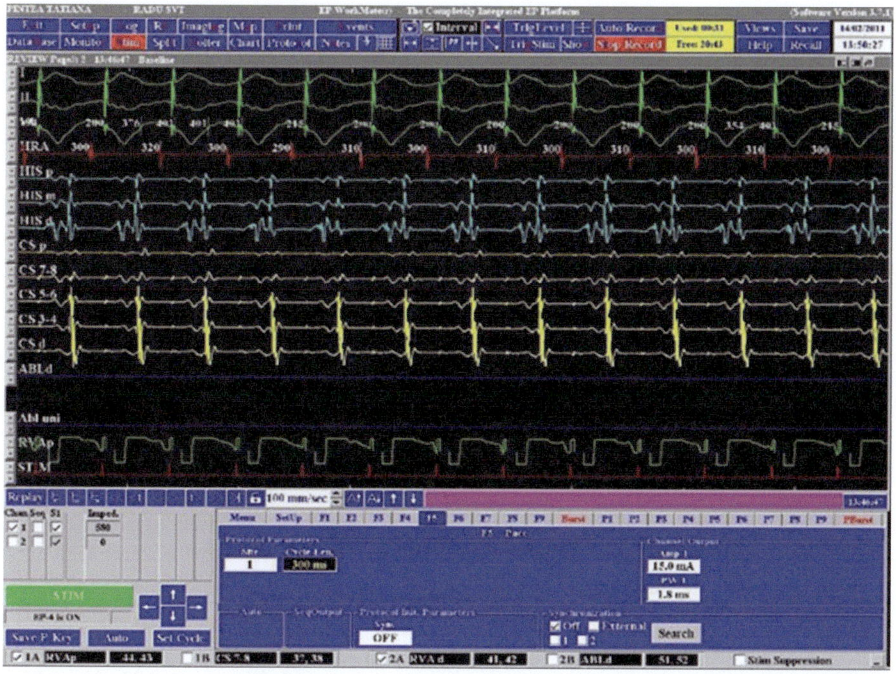

Fig. 21.4 One-to-one retrograde conduction during right ventricular pacing. Please note that after each spike of stimulation (STIM), there is one ventricular activation (RVAp, green color) and one atrial activation on the right atrium, His, or coronary sinus electrogram (HRA, red color, or His, blue color, or CS, yellow color)

independently of the underlying disease: AV block or sinus node disease (even in patients with infrahisian block, there was an increase in the ventricular rate). The mean dose used by the authors was 0.1 μg/kg/min. There were no significant side effects: the blood pressure increased to a maximum of 180/100 mmHg. The only side effect mentioned was tremor, which was well supported by the patients and disappeared after stopping the infusion.

21.5 Assessment of Retrograde Ventriculoatrial Conduction

VA conduction is very easy to assess during an electrophysiological study: we stimulate the right ventricular apex, and we look for the atrial activation on the HRA catheter or His catheter or coronary sinus catheter (Fig. 21.5).

During pacemaker implantation, assessment of VA conduction can be done using either intracardiac electrograms or surface ECG markers: in the case of patients with double-chamber pacemakers, after the implantation of both leads inside the right ventricle and right atrium, retrograde conduction can be tested by stimulating the right ventricle and verifying for atrial activation after each ventricular capture

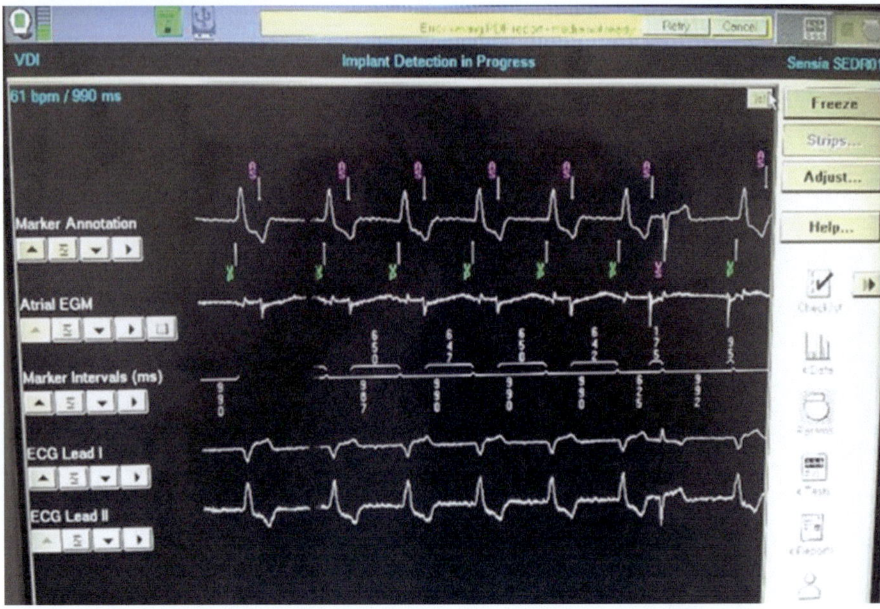

Fig. 21.5 Assessment of ventriculoatrial conduction in a patient with double-chamber pacemaker. Please note that after each ventricular-paced (VP) beat, there is one atrial-sensed (AS) beat. There is one ventricular-sensed (VS) beat without any atrial-sensed (AS) beat because probably the retrograde atrium is in the PVARP period

(Fig. 21.6). In the case of monochamber pacemaker implantation, there are two possibilities:

1. Look for a retrograde P wave using amplified (Fig. 21.7) ECG.
2. Insert a supplementary electrode inside the right atrium (either a pacing electrode or an EP electrode to assess for atrial activation, Fig. 21.8).

When 12-lead ECG is used for P wave assessment, settings can be modified for an amplitude of 20 mm/1 mV instead of 10 mm/1 mV. In the case of VVI single-chamber pacemakers, a supplementary EP catheter can be inserted inside the right atrium or coronary sinus through the subclavian vein or internal jugular vein which is further connected to the Medtronic/Saint Jude/Biotronik programmer or to an ECG machine with the gain set to maximum of signal amplitude. Ventricular pacing should be initiated faster than the heart rate increased by adrenaline: 5–10 bpm above the intrinsic rate (e.g., if the patient has a heart rate of 40 bpm during complete AV block which further increases to 65 bpm after adrenaline infusion, ventricular pacing should be initiated to at least 70 bpm to assess VA conduction). Intracardiac electrical stimulation can be performed using the programmable

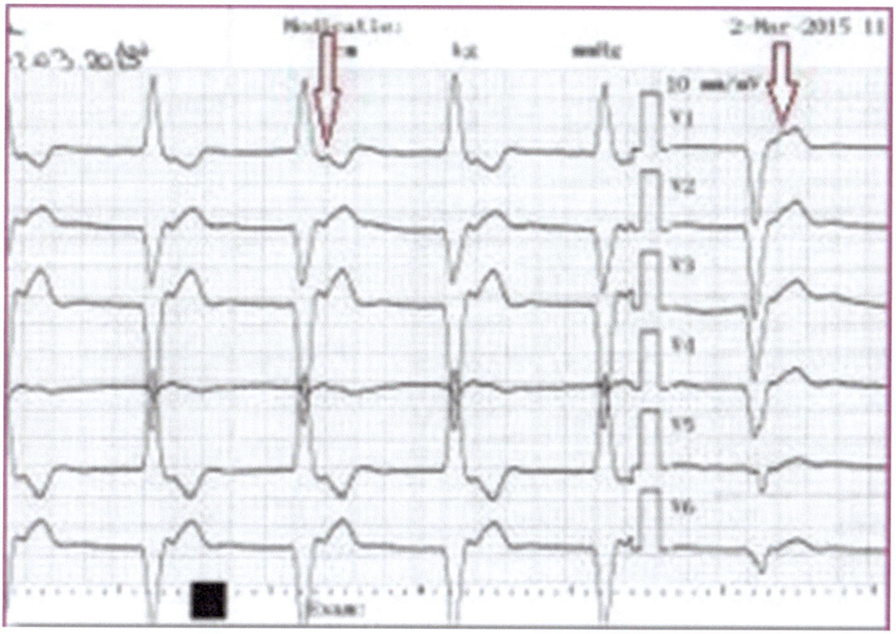

Fig. 21.6 Assessment of ventriculoatrial conduction on a 12-lead ECG after right ventricular stimulation at a rate of 60 bpm. Please note the retrograde P wave at the end of the QRS complex (red arrows)

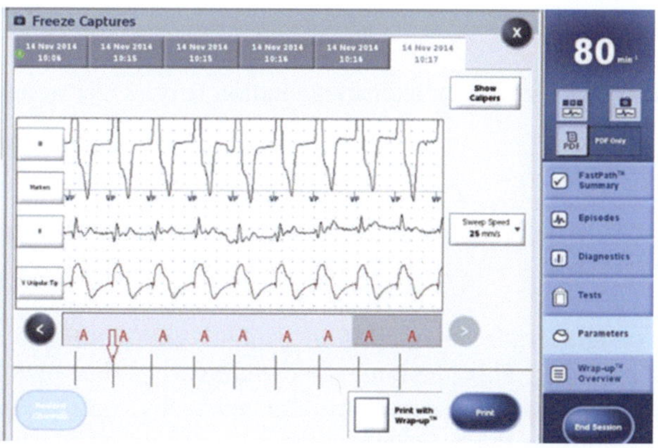

Fig. 21.7 Assessment of ventriculoatrial conduction during implantation of a Saint Jude ventricular pacing lead. Please note on the same image was added the atrial electrogram using a supplementary EP electrode at the level of high right atrium. After each ventricular-paced beat at a rate of 80 bpm, there is an atrial electrogram, confirming the presence of retrograde VA conduction

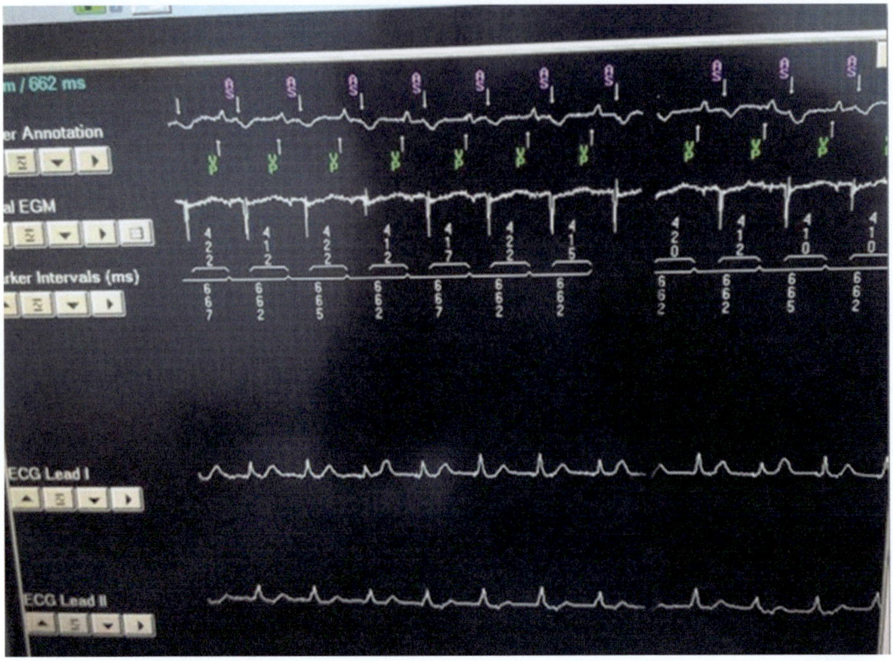

Fig. 21.8 Assessment of ventriculoatrial conduction after 0.1 μg/kg/min infusion of adrenaline. As the heart rate of this patient with 2:1 AV block increased to 85 bpm, the stimulation was performed at a rate of 90 bpm after implantation of a double-chamber pacemaker. Please note after each ventricular-paced (VP) beat, there is one atrial-sensed (AS) beat

stimulator that is used for pacemaker implantation (Fig. 21.9) or an EP recording system.

21.6 Clinical Implication of Retrograde Ventriculoatrial Conduction

The presence of retrograde conduction in patients with complete antegrade AV block is not so rare [10]. This phenomenon explains why some patients with complete AV block can present pacemaker-mediated tachycardia despite an absent antegrade conduction to the ventricles [11]. Several studies have found that VA conduction can be present in up to 25% of patients with complete AV block [12, 13]. When VA conduction is present, most probably the site of antegrade conduction block is infrahisian. Josephson et al. reported similar results: if VA conduction is absent and retrograde VA conduction is present, the site of AV block is infrahisian [14].

Fig. 21.9 Medtronic programmer that can be used to test pacemaker-mediated tachycardia

21.6.1 Pacemaker Syndrome

Pacemaker syndrome can be clinically manifested when there is retrograde VA conduction and the atrial activation interferes with the ventricular electrical and mechanical activation. The clinical manifestations of the pacemaker syndrome are dizziness, lightheadedness, fatigue, hypotension, and congestive heart failure [13]. The presence of pacemaker syndrome is up to 20% in patients with sick sinus syndrome and up to 13.2% in patients with antegrade AV block [14]. In the study of Akhtar, ECG showed retrograde VA conduction in 25 of 121 cases [15].

In the study of Adams et al., pacemaker syndrome was present in 12.4% of patients who had complaints: weakness, fatigability, slow thinking, dizziness, dyspnea during an effort, angina at rest or during exertion, pulsation of neck veins, hypotension, or syncope. Retrograde P wave was discovered on standard ECG leads in 80.9% of the patients with these complaints and only on transesophageal ECG in 19% [10].

21.6.2 Pacemaker-Mediated Tachycardia

Retrograde VA conduction may predispose to pacemaker-mediated tachycardia in patients with double-chamber pacing. Although VA time increases with most of the antiarrhythmic drugs, verapamil [16, 17], metoprolol, atenolol, amiodarone [18], propafenone, and flecainide [19], it is infrequently eliminated [12].

For pacemaker-mediated tachycardia induction, retrograde atrial activation should appear after the termination of post-ventricular atrial refractory period (PVARP). VA conduction time has different values in function of the type of AV block. Using transesophageal ECG recording in 79 patients were divided in three groups: normal, prolonged AV conduction time without bundle branch block, and prolonged AV conduction time with bundle branch block. Longest VA conduction time was found in the group of atrioventricular conduction disturbance, with bundle branch block. This category of patients is more prone to present a VA time, with P wave outside the PVARP [11].

There are two types of pacemaker-mediated tachycardias. Both of them can be facilitated by adrenaline infusion, if retrograde VA conduction can be induced. The two types of PMTs are ELT or endless loop tachycardia [20, 21] and RNRVAS or repetitive nonreentrant VA synchrony [22, 23]. Both endless loop tachycardia and repetitive nonreentrant VA synchrony depend on the presence of retrograde ventriculoatrial conduction and have similar starting and terminating mechanisms [24]. Endless loop tachycardia occurs when the atrial channel senses a retrograde P wave and starts a reentrant loop using the pacemaker as the antegrade limb and the conduction system as the retrograde limb of the reentry circuit [25]. RNRVAS also occurs during dual-chamber pacing if a ventricular-paced beat induces a retrograde-conducted P wave that remains unsensed by the atrial channel because of the PVARP. Pacemaker-mediated tachycardia persists until ventriculoatrial conduction stops or the retrograde atrial activation is not sensed by the atrial channel. The presence of ventriculoatrial retrograde conduction may vary in the function of the autonomic tone as well as the rate of pacemaker-mediated tachycardia [26].

In older types of pacemakers, ELT and RNRVAS could occur because they did not have the capability to program an atrial refractory period. In newer devices a post-ventricular atrial refractory period can be programmed and thus prevent pacemaker-mediated tachycardias.

21.6.3 Defibrillator Discrimination Algorithm

In the case of bundle branch block or aberrancy, the discrimination algorithm between supraventricular and ventricular tachycardia may fail. For double-chamber devices, the presence of a supplementary lead inside the left atrium can further discriminate between tachycardias with retrograde VA dissociation and tachycardias with VA 1:1 conduction. Usually ventricular tachycardia manifests with VA dissociation. The problem of discrimination appears when VA conduction is present. In order to verify its presence during physiological or pathophysiological stresses, we can infuse adrenaline 0.05–0.1 μg/kg/min in such patients and test for retrograde conduction. Both parasympathetic and sympathetic nervous systems can influence ventriculoatrial conduction in patients with defibrillators, as demonstrated by the study of Sadr-Ameli and Klementowicz [5, 27].

References

1. Westveer DC, Stewart JR, Goodfleish R, Gordon S, Timmis GC. Prevalence and significance of ventriculoatrial conduction. Pacing Clin Electrophysiol. 1984;7:784–9.
2. Akhtar M, Damato AN, Batsford W, Ruskin JN. A comparative analysis of anterograde and retrograde conduction patterns in man. Circulation. 1975;52:766–78.
3. Shenasa M, Gilbert CJ, Schmidt DH, Akhtar M. Procainamide and retrograde atrioventricular nodal conduction in man. Circulation. 1982;65:355–62.
4. Richter S, Muessigbrodt A, Salmas J, Doering M, Wetzel M, Arya A, et al. Ventriculoatrial conduction and related pacemaker-mediated arrhythmias in patients implanted for atrioventricular block: an old problem revisited. Int J Cardiol. 2013;168:3300–8.
5. Sadr-Ameli MA, Shenasa M, Lacombe P, Faugere G, Nadeau R. Effect of autonomic nervous system modulation on retrograde atrioventricular nodal conduction in the human heart. Cardiovasc Res. 1987;21:45–54.
6. Hariman RJ, Pasquariello JL, Gomes JA, Holtzman R, El-Sherif N. Autonomic dependence of ventriculoatrial conduction. Am J Cardiol. 1985;56:285–91.
7. Morady F, Nelson SD, Kou WH, Pratley R, Schmaltz S, De Buttleir M, et al. Electrophysiologic effects of epinephrine in humans. J Am Coll Cardiol. 1988;11:1235–44.
8. Morady F, Kou WH, Kadish AH, Toivonen LK, Kushner JA, Schmaltz A. Epinephrine-induced reversal of verapamil's electrophysiologic and therapeutic effects in patients with paroxysmal supraventricular tachycardia. Circulation. 1989;79:783–90.
9. Cismaru G, Gusetu G, Muresan L, Rosu R, Andronache M, Matuz R, et al. Recovery of ventriculo-atrial conduction after adrenaline in patients implanted with pacemakers. Pacing Clin Electrophysiol. 2015;38:857–63.
10. Adams CW. Retrograde atrial conduction with complete heart block following implantation of an internal ventricular pacemaker. Dis Chest. 1963;43:544–5.
11. Castillo C, Samet P. Retrograde conduction in complete heart block. Br Heart J. 1967;29:553–8.
12. Hayes DL, Furman S. Atrio-ventricular and ventriculo-atrial conduction times in patients undergoing pacemaker implant. Pacing Clin Electrophysiol. 1983;6:38–46.
13. Touboul P, Huerta F, Delahaye JP. Retrograde conduction in complete atrioventricular block. Study using His bundle recordings. Br Heart J. 1976;38:706–11.
14. Josephson ME. Clinical cardiac electrophysiology: techniques and interpretation. 4th ed. Philadelphia: Lippincott Williams and Wilkins; 2008. p. 106–7.

15. Akhtar M. Retrograde conduction in man. PACE. 1981;4:548–62.
16. Kashani IA, Shakibi JG, Siassi B. Effect of verapamil on retrograde ventriculoatrial conduction in children. Jpn Heart J. 1980;21:485–90.
17. Shanasa M, Denker S, Mahmud R, Lehmann M, Murthy VS, Akhtar M. Effect of verapamil on retrograde atrioventricular nodal conduction in the human heart. J Am Coll Cardiol. 1983;2:545–50.
18. Reddy CP, Kuo CS. Effect of amiodarone on retrograde conduction and refractoriness of the His-Purkinje system in man. Br Heart J. 1984;51:648–53.
19. Dougherty AH, Rinkenberger RL, Naccarelli GV. Effect of pharmacologic autonomic blockade on ventriculoatrial conduction. Am J Cardiol. 1986;57:1274–9.
20. Den Dulk K, Lindemans FW, Bar FW, Wellens HJ. Pacemaker related tachycardias. Pacing Clin Electrophysiol. 1982;5:471–5.
21. Bathen J, Gundersen T, Forfang K. Tachycardias related to atrial synchronous ventricular pacing. Pacing Clin Electrophysiol. 1982;5:471–5.
22. Barold SS, Levine PA. Pacemaker repetitive nonreentrant ventriculoatrial synchronous rhythm. A review. J Interv Card Electrophysiol. 2001;5:45–58.
23. Barold SS. Repetitive reentrant and nonreentrant ventriculoatrial synchrony in dual chamber pacing. Clin Cardiol. 1991;14:754–63.
24. Bathen J, Gundersen T, Forfang K. Tachycardia related to atrial synchronous ventricular pacing. PACE. 1982;5:471–5.
25. Frumin H, Furman S. Endless loop tachycardia started by an atrial premature complex in a patient with dual chamber pacemaker. J Am Coll Cardiol. 1985;5:707–10.
26. Ausubel K, Gabry MD, Klementowicz PT, Furman S. Pacemaker-mediated endless loop tachycardia at rates below the upper rate limit. Am J Cardiol. 1988;61:465–7.
27. Klementowicz P, Ausubel K, Furman S. The dynamic nature of ventriculoatrial conduction. PACE. 1988;11:1917–28.

What to Do When Clinical Arrhythmia Is Uninducible?: Stepwise Approach

22

Gabriel Cismaru

G. Cismaru (✉)
"Iuliu Hatieganu" University of Medicine and Pharmacy, 5th Department of Internal Medicine, Cardiology-Rehabilitation, Cluj-Napoca, Romania
e-mail: cismaru.gabriel@umfcluj.ro

G. Cismaru (ed.), *Arrhythmia Induction in the EP Lab*,
https://doi.org/10.1007/978-3-319-92729-9_22

22.1 Introduction

Catheter ablation has been established as a versatile modality to treat a wide range of tachyarrhythmias. Ventricular tachycardia ablation is performed after mapping during sinus rhythm or during tachycardia. Cost-effectiveness studies have generally favored catheter ablation over medical treatment because of its potential curative effect. In the electrophysiology (EP) laboratory, induction and maintenance of clinical arrhythmias is the cornerstone of diagnosis and treatment, especially in cases where the clinical arrhythmia is not documented on baseline electrocardiogram. Activation mapping during VT is felt to be the most reliable method for the identification of the ablation target but, likewise, requires the VT to be inducible. During VT, a three-dimensional mapping system permits the reconstruction of the reentry circuit, allowing subsequent RF application at the level of reentry isthmus. However, VT induction is not always possible in the basal state. With this, a various number of medications were started to be used as a means to facilitate arrhythmia induction, including isoprenaline, adrenaline, and atropine. Choi et al. demonstrated that catheter ablation for sustained outflow tract VT yielded better results when the arrhythmia was inducible [1]. Attempts to induce VT are important in this specific population, of which aminophylline and isoprenaline have been used for induction. With this, it has been established that the most common mechanism of idiopathic outflow tract VT is cyclic AMP-mediated triggered activity.

The induction of ventricular tachycardia is highly significant in patients with ischemic cardiomyopathy (ICM). Both fast VT and slow VT have demonstrated adverse long-term outcome in patients with ICM. When VT is non-inducible, drugs can be used to facilitate arrhythmia induction. Most published studies were made using isoproterenol [2]. Thus, in essence, ventricular tachycardia induction is useful for:

1. Substrate mapping and activation mapping
2. Verification of the efficacy of after VT ablation
3. Working up various differential diagnoses of a wide complex QRS tachycardia, such as SVT fifth bundle branch block/antegrade activation through an accessory pathway/VT

22.2 Drugs That Can Be Injected as a Bolus

It has been noted that most EP specialists prefer to rapidly inject a bolus of a drug for arrhythmia induction. Infusions takes a longer period of time, with errors in dilution and preparation sometimes noted. Therefore in some instances, a bolus might be preferred. However, we do not recommend boluses of adrenaline, although there are studies, such as that of Choi et al., where a bolus of 20–100 μg was injected for PVC/VT induction from RVOT, Our experience demonstrated an increased risk of high blood pressure and subsequent bouts of headache. It is well known that adrenaline can induce a potentially increase in the blood pressure with resultant cerebral hemorrhage or acute pulmonary edema. In our EP lab, there were instances where extreme hypertension (>230/120 mmHg) was noted even after a small dose of 50 μg of adrenaline was administered, followed by persistent headache >48 h. This particular patient was worked up with an MRI of the brain demonstrating no lesions. However, due to the potentially catastrophic symptoms, we had opted to avoid the use of the bolus protocols. Other beta-mimetics like noradrenaline, dopamine, and dobutamine may potentially have similar effects; thus such medications are likewise given as infusions.

However, in existing literature, boluses of isoprenaline, adenosine, atropine, and ephedrine are injected for arrhythmia induction in adults or in children, without any safety concerns. The authors therefore report their own experiences with such regimens in the presence of these studies.

22.3 Drugs That Can Be Infused

As previously discussed, the risk of a potentially sudden increase in the blood pressure makes adrenaline, noradrenaline, dopamine, and dobutamine less suitable for bolus administration. In these cases, an infusion can be started after a first basal attempt to induce arrhythmia, with programmed stimulation repeated during and after the infusion. The disadvantage is the time taken for preparation and injection of the infusion, but this is offset by the quick termination of drug effects if administered in this particular manner.

There is no argument for infusing adenosine or atropine as there is no benefit over the bolus administration.

22.4 Drugs Used for Arrythmias That Occur During Physiological or Pathophysiological Stresses

In several patients, clinical arrhythmias occur during particular forms of physiological or pathophysiological stress. It is well demonstrated that a concentration of 862 pg/mL of adrenaline is seen in the bloodstream during several forms of physiological or pathophysiological stresses: smoking, public speaking, mild hypoglycemia, submaximal stress test, dental removal, or minor surgery. A higher concentration of 1.374 pg/mL occurs in maximal stress test, myocardial infarction, severe hypoglycemia, and diabetic ketoacidosis.

Most of the drugs presented in this book are beta-mimetics and will increase the serum concentration of adrenaline and noradrenaline: isoprenaline, adrenaline, orciprenaline, ephedrine, noradrenaline, dopamine, and dobutamine. By stimulating the β1 receptors, they exert both inotropic and chronotropic effects. Furthermore, stimulation of β2 receptors may induce further chronotropic effects by an indirect reflex mediated by baroreceptors and initiated by peripheral vasodilation leading to norepinephrine release in the bloodstream (Figs. 22.1, 22.2, and 22.3).

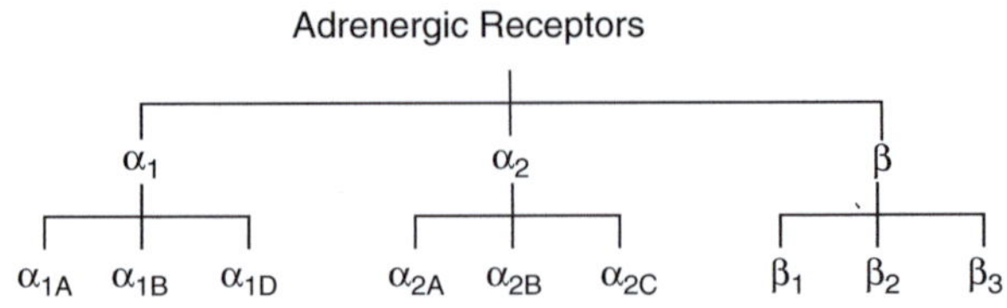

Fig. 22.1 Types of adrenergic receptors

Fig. 22.2 Classification of adrenergic hormone receptors. *A* adrenaline, *NA* noradrenaline

Receptor	Agonists
alpha$_1$ (α_1)	NA>A
alpha$_2$ (α_2)	A>NA
beta$_1$ (β_1)	A=NA
beta$_2$ (β_2)	A >>NA

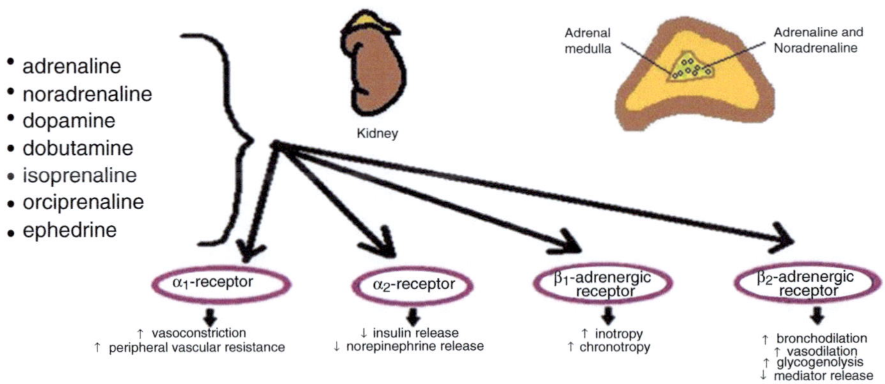

Fig. 22.3 Beta-adrenergic drugs act on alpha- and beta-receptors in the body with several effects

The main action of salbutamol is the β2 stimulation effect. A β2 adrenergic effect at the level of the sinus node was demonstrated but with a minor effect on cardiac chronotropism [3, 4]. The β2 direct chronotropic and inotropic effects were earlier demonstrated by infusion of terbutaline, a medication with β2 selectivity that induced increases in the heart rate and blood pressure. This increase in the heart rate and systolic blood pressure could not be prevented by atenolol, a β1 blocker, with and without atropine added [5, 6]. β2 receptors are found in the atrial tissue in a percentage of 20–40% of the overall number of adrenergic receptors and at the level of ventricular myocardium at 10–40%. Compared to the myocardium, the percentage of β2 receptors is higher in the AV node [7].

In the study by Rodefeld et al., the density of β receptors was found to be three times higher in the AV node compared to the atrial myocardium. β2 receptors also represent 30% of the totality of sinus node receptors [8]. The cardiac effects of β2 adrenoreceptors include the following: sinus tachycardia and ventricular and atrial premature contractions, as well as ventricular and atrial rapid arrhythmias [9, 10]. The increase of AV node conduction results in faster ventricular response during atrial fibrillation or atrial flutter.

Therefore, for patients with arrhythmias that are known to occur during several types of stresses, beta-mimetics can be used in the EP lab for arrhythmia induction. It is important for those patients to stop medication with beta-blockers at least 48 h before performing the electrophysiological testing.

22.5 Ten-Step Approach to Arrythmia Induction in the EP Lab

As part of our observations throughout the years, the authors propose the following ten-step protocol for arrhythmia induction in the EP laboratory:

- *Step 1*. **Stop medication.** For patients with supraventricular or ventricular tachycardia that needs an ablation, the first, but often overlooked measure, is to stop medication at least five half-lives before the EP study. A thorough drug history assessment and checklist for specific medications may be useful in ensuring proper compliance and timing.
- *Step 2*. **Red Bull**. The second step is to give two cans of Red Bull, preferably 1 h before the electrophysiological study, in order to permit the caffeine to be fully absorbed by the time of intervention. This further increases the heart rates of patients and may be arrhythmogenic.
- *Step 3*. **Programmed atrial and ventricular stimulation**. During EP studies, we try to induce SVT or VT by programmed atrial and ventricular stimulation, by up to three extrastimuli on an imposed rhythm of 600 and 400 ms. Stimulation of the coronary sinus and the institution of a fourth extrastimulus may be done.
- *Step 4*. **Isoproterenol.** If no arrhythmia is induced in the basal state, an infusion of isoprenaline for arrhythmia induction may help. Orciprenaline is used in some European countries instead of isoprenaline.
- *Step 5*. **Adrenaline.** In the case of lack of isoprenaline, we have found that infusion of adrenaline is also effective, although it has been shown to take more time than isoprenaline to achieve a steady state.
- *Step 6*. **Ephedrine.** If a bolus injection is preferred over a continuous infusion of adrenaline, we give a bolus of ephedrine. If the patient has already received infusion of adrenaline without arrhythmia induction, do not add an epinephrine bolus, as this will further increase the blood pressure. Instead, it is preferable to add a bolus of atropine.
- *Step 7*. **Atropine**. As earlier discussed, in cases of non-inducibility under adrenaline, a dose of atropine of 1 mg can be added over adrenaline.
- *Step 8*. **Caffeine, theophylline, or salbutamol**. For patients with PVC ablation, caffeine, theophylline, and salbutamol can be attempted in cases of non-inducibility.

- *Step 9.* **Adenosine**. For patients with AVNRT, low-dose adenosine can be used for echo beat and PSVT induction. For patients with atrial fibrillation, adenosine bolus or infusion can be attempted to unmask dormant pulmonary vein conduction.
- *Step 10.* **Dopamine, dobutamine, and noradrenaline** should be used in selected cases, when the clinical arrhythmia is resistant and non-inducible with other drugs (like monomorphic PVCs) or when drug availability remains a problem.

Among the various steps in the protocol described, the most novel approach to arrhythmia induction has been noted to be our use of Red Bull. We usually give two cans of Red Bull, 1 h before the electrophysiological study in patients with ventricular premature contractions or ventricular tachycardia when we want to perform activation mapping. With this, we describe two case reports (Case No. 1 and Case No. 2) where arrhythmia induction was achieved with Red Bull. The third presentation reports the case of a football referee that received adrenaline infusion for arthythmia induction.

22.6 Case Presentations

22.6.1 Case No. 1

A 36-year-old female patient with palpitations, a burden of 35% on 24 h of PVCs, and repeated episodes of monomorphic ventricular tachycardia (VT) was referred for catheter ablation. During three episodes of sustained VT, she presented syncope because of hemodynamic compromise with a heart rate of 240 bpm. PVCs and VT had an RVOT morphology with left bundle branch block pattern in lead V1 and inferior axis and precordial transition in lead V4. Metoprolol and verapamil had no effect on the PVC burden, so the patient was treated with amiodarone. Despite amiodarone administration, the patient still presented with a high number of PVCs 29% on 24 h; therefore catheter ablation was proposed. Fourteen days prior to ablation, amiodarone was stopped. On the day of the procedure, the patient had no PVCs and was noted to be asymptomatic. Prior to bringing the patient to the EP room, she presented with no PVCs on the 12-lead ECG. After a period of continuous monitoring, venous puncture, and placement of the catheters inside the heart chambers, no PVCs were noted to recur. Therefore the patient received two cans of Red Bull, and after 15 min she presented a high number of PVCs and repetitive episodes of VT (Fig. 22.4). The morphology was similar to the clinical PVC/VT. Activation mapping was performed with the use of EnSite-NavX system and the use of a 4-mm tip Saint Jude ablation catheter placed in the RVOT. The origin of the PVCs was found by point-by-point mapping in the anteroseptal region of the RVOT, where a −40 ms potential was present before the onset of QRS. Two RF applications were performed at 35 W with temperatures under 55 °C with a duration of 60 s. No PVCs were present afterward. Holter ECG revealed 128 atrial premature contractions without any PVCs or VT. At 2 months follow-up, the patient presented with no palpitations or syncopal attacks, and Holter ECG showed 0 PVCS on 24 h. She remained without any medication.

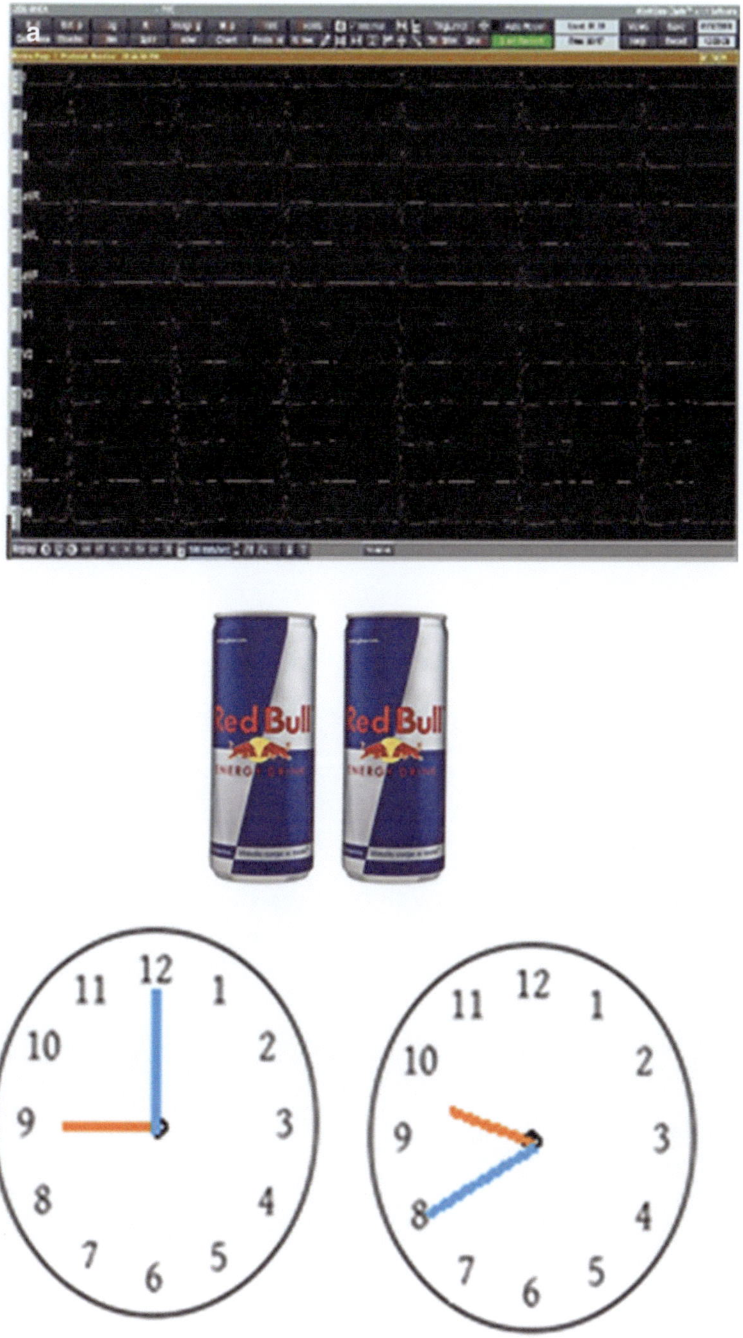

Fig. 22.4 (**a**) Before Red Bull the patient had no PCVs, no VT. (**b**) After Red Bull, at 40-min intervals, the patient presented multiple PVCs and (**c**) episodes of ventricular tachycardia of 6–7 ventricular complexes

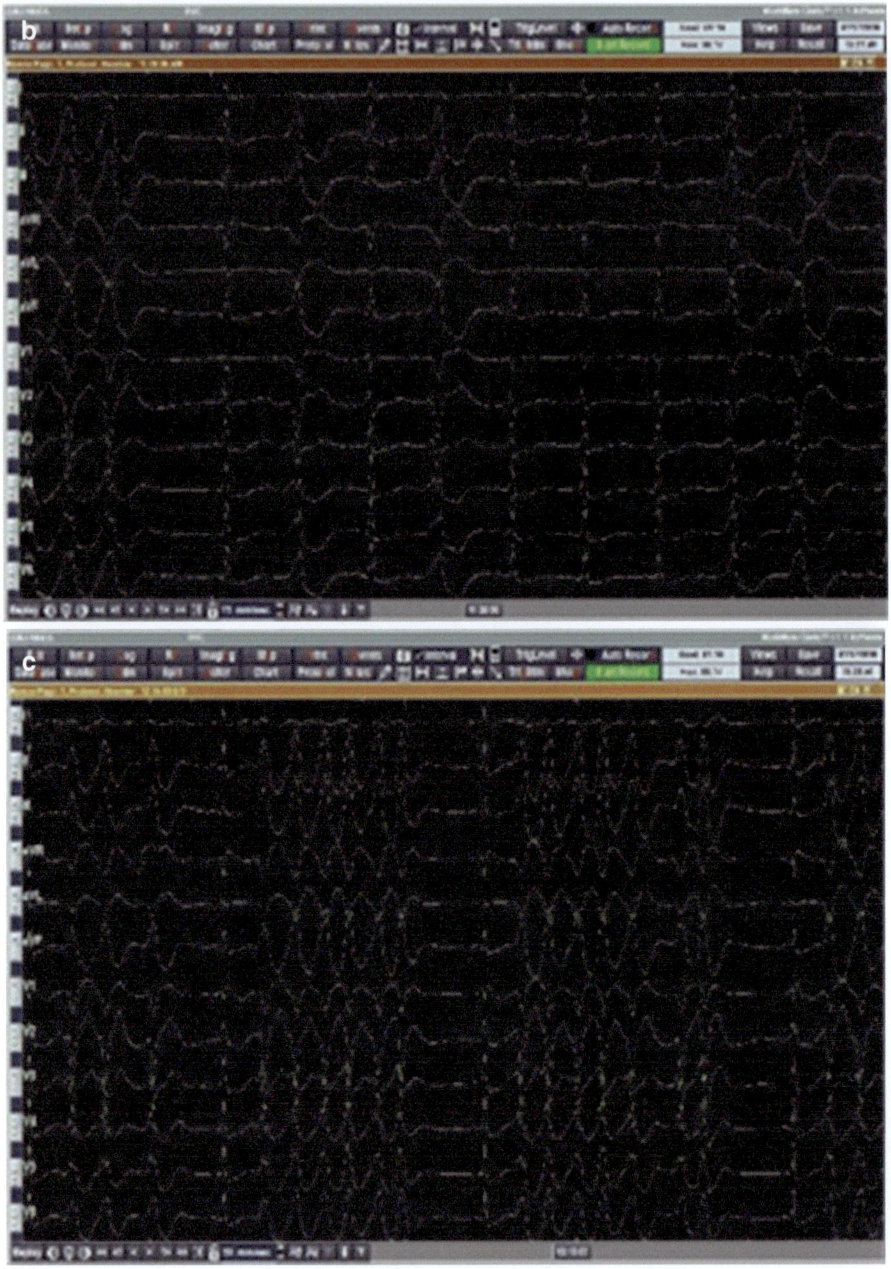

Fig. 22.4 (continued)

22.6.2 Case No. 2

A 63-year-old patient with old apical infarction was referred to our center for catheter ablation of ventricular tachycardia. The patient had ischemic heart disease and underwent double coronary artery bypass had an old apical infarction and an ejection fraction of 30%. Because he presented three episodes of VT in 6 months, he was referred for catheter ablation.

When the patient came to our cardiology department, amiodarone was stopped 60 days before the EP study. ICD interrogation showed no VT during the last 6 weeks. On the day of the EP study at 7:30, the patient received two cans of Red Bull, and starting 8:30 (Figs. 22.5, 22.6, and 22.7) until 10:00, he presented with four episodes of ventricular tachycardia, unresponsive to ATP, that were terminated by internal electrical shock.

During three-dimensional mapping, an apical zone was found to be responsible for VT. Catheter ablation at this level during VT stopped the arrhythmia. At the end of EP study, no VT was induced after adrenaline infusion and ephedrine injection. At 9 months there were no recurrences of VT.

Fig. 22.5 At 7:30 the patient received two cans of Red Bull

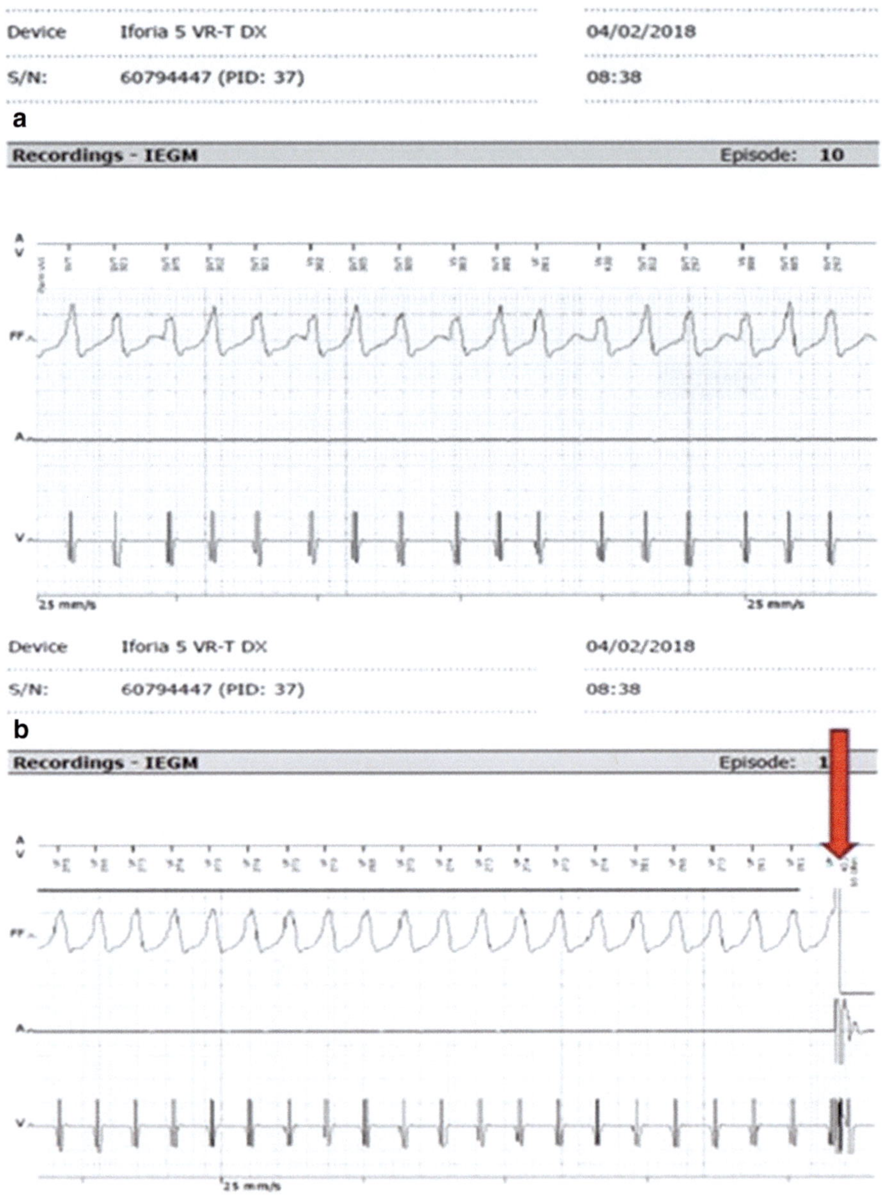

Fig. 22.6 At 8:30 the patient presented VT, which necessitated an electrical shock (red arrow). During the following 90 min, he presented another three episodes of ventricular tachycardia that necessitated internal cardioversion

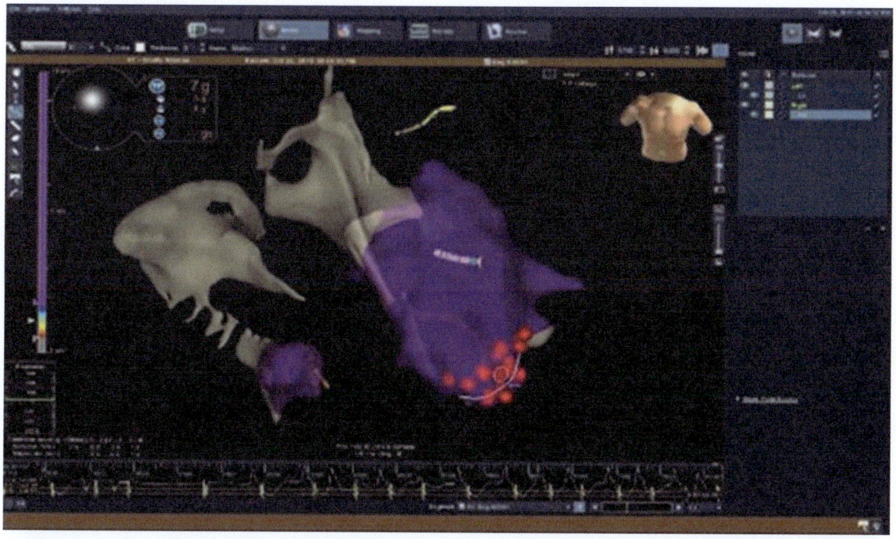

Fig. 22.7 Catheter ablation of an apical ventricular tachycardia. At the apex of the left ventricle, there are many red dots which are ablation points. Each time the catheter reached this zone, ventricular tachycardia was mechanically induced. Catheter ablation at this level made VT uninducible with and without adrenaline

22.6.3 Case No. 3

A 32-year-old Romanian second-league referee had paroxysmal episodes of palpitations that started 1 year before his presentation at the cardiology department. His ECG was normal, and echocardiography showed normal values. For arrhythmia detection, several Holter ECGs failed to show abnormalities as well as cardiac stress test conducted to 275 W.

He monitored his heart rate using a Polar device during training in order to detect the clinical arrhythmia. Successive recordings showed increase in the heart rate to 195 bpm during exercise, without episodes of paroxysmal palpitations. One day during "yo-yo" intermittent test, he presented an unexpected episode of palpitations with sudden increase of heart rate from 160 to 220 bpm that lasted for 5 min confirming a paroxysmal episode of tachycardia (Figs. 22.8 and 22.9).

We performed an electrophysiological study that revealed dual nodal conduction. No arrhythmia was induced after 1 mg bolus of atropine, and the conduction jump of 50 ms disappeared after atropine. Infusion of 0.05 μg/kg adrenaline with programmed stimulation facilitated induction of a typical slow-fast reentrant tachycardia at 200/min. After successful slow pathway ablation, at 36 months follow-up, the patient had no more episodes of palpitations during rest or effort (Figs. 22.10, 22.11, and 22.12).

Fig. 22.8 The Polar device used for monitoring the heart rate

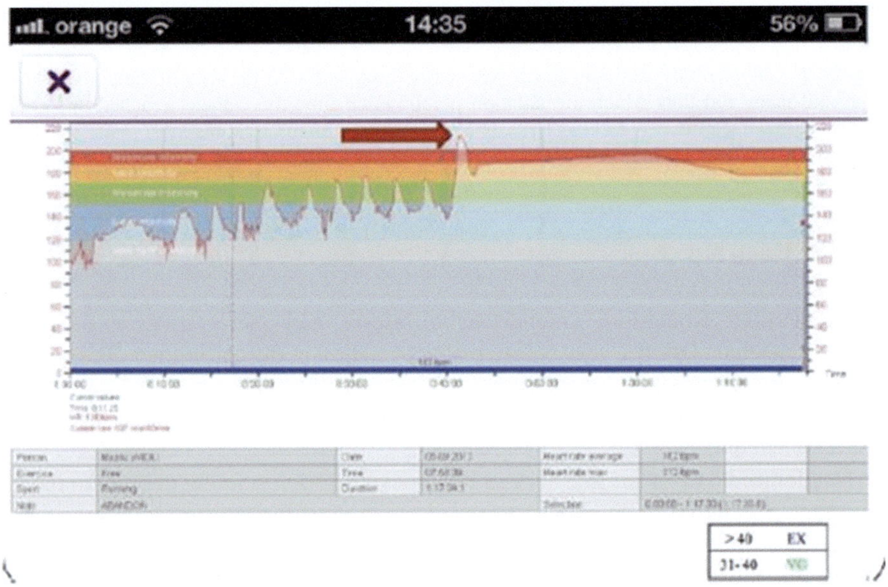

Fig. 22.9 Heart rate diagram during training shows a brutal onset episode of tachycardia 215 bpm. The patient felt palpitations more severely than normal palpitations that he presented during a maximal exercise

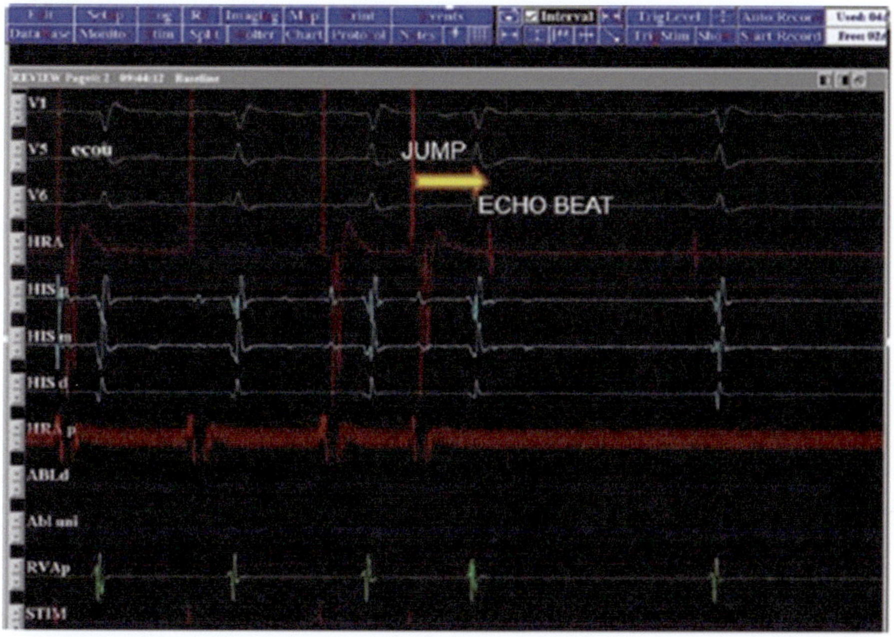

Fig. 22.10 Before atropine bolus the patient had a jump of >50 ms demonstrating the presence of dual nodal pathway

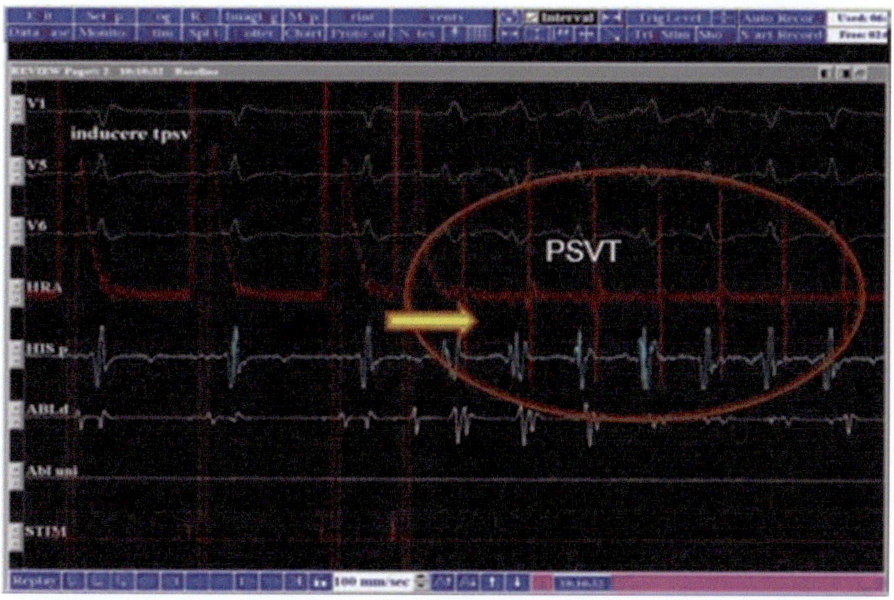

Fig. 22.11 After adrenaline infusion the patient presented paroxysmal supraventricular tachycardia

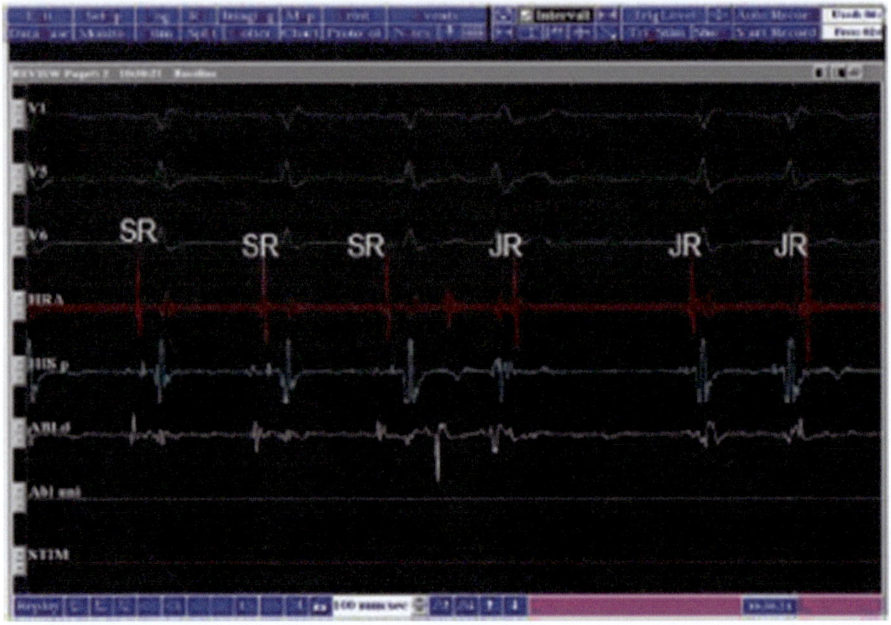

Fig. 22.12 During catheter ablation, junctional rhythm was present, a sign of ablation success

References

1. Choi EK, Kumar S, Nagashima K, Lin KY, Barbhaiya CR, Chinitz JS, et al. Better outcome of ablation for sustained outflow-tract ventricular tachycardia when tachycardia is inducible. Europace. 2015;17:1571–9.
2. Niebauer M, Daoud E, Goyal R, Chan KK, Harvey M, Bogun F, et al. Use of isoproterenol during programmed ventricular stimulation in patients with coronary artery disease and non-sustained ventricular tachycardia. Am Heart J. 1996;131:516–8.
3. Simonson BG, Stiska J, Strom B. Double-blind trial with increasing doses of salbutamol and terbutaline aerosols in patients with reversible airway obstruction. Acta Med Scand. 1972;192:371–6.
4. Gray BF, Frame M, Costello JF. A comparative double blind study of the bronchodilator effects and side effects of inhaled fenoterol and terbutaline administered in equipotent doses. Br J Dis Chest. 1982;76:341–50.
5. Strauss MH, Reeves RA, Smith DL, Leene FHH. The role of cardiac β-1 receptors in the hemodynamic response to beta-2 agonist. Clin Pharmacol Ther. 1986;40:108–15.
6. Levine MAH, Leenen FHH. Role of beta 1 receptors and vagal tone in cardiac inotropic and chronotropic responses to a beta 2 agonist in humans. Circulation. 1989;79:107–15.
7. Summers RJ, Molnaar P, Russel F, Elnatan J, Jones CR, Buxton BF, et al. Coexistence and localization of beta 1 and beta 2 adrenoreceptors in the human heart. Eur Heart J. 1989;10:11–21.
8. Rodefeld MD, Beau SL, Schuessler RB, Boineau JP, Saffitz JE. Beta adrenergic and muscarinic cholinergic receptor densities in the human sinoatrial node. J Cardiovasc Electrophysiol. 1996;7:1039–49.
9. Crane J, Burgess C, Beasley R. Cardiovascular and hypokalemic effects of inhaled salbutamol, fenoterol, and isoprenaline. Thorax. 1989;44:136–40.
10. Flatt A, Crane J, Purdie G, Kwong T, Beasley R, Burgess C. The cardiovascular effects of betaadrenergic agonist drugs administered by nebulization. Postgrad Med J. 1990;66:98–101.

MIX
Papier aus verantwortungsvollen Quellen
Paper from responsible sources
FSC® C105338

If you have any concerns about our products,
you can contact us on
ProductSafety@springernature.com

In case Publisher is established outside the EU,
the EU authorized representative is:
Springer Nature Customer Service Center GmbH
Europaplatz 3, 69115 Heidelberg, Germany

Printed by Libri Plureos GmbH
in Hamburg, Germany